SHAPING
THE
FUTURE

SHAPING
THE
FUTURE

Business Design through Information Technology

PETER G.W. KEEN

 Harvard Business School Press

99 98 97 96   10 9 8 7 6 5

The recycled paper used in this publication meets the requirements of
the American National Standard for Permanence of Paper for Printed Library
Materials Z39.49-1984.

Library of Congress Cataloging-in-Publication Data

Keen, Peter G. W.
   Shaping the future : business design through information
technology / Peter G.W. Keen.
      p.   cm.
   Includes bibliographical references and index.
   ISBN 0-87584-237-2 (hardcover : acid free paper)
   1. Business—Data processing—Management.   2. Business—
Communication systems—Management.   3. Management infoi   ation
systems.   4. Organizational change—Management.   I. Title.
HF5548.2.K395   1991
658.4'038—dc20   90-49732
                                                          CIP

For Lynda

My partner, best critic,
and practical futurist.

My name is on the title page,
but your contribution and influence
are on every page.

Contents

Preface and Acknowledgments

My first academic appointment was in the Organizational Studies group at the Sloan School of Management. My last full-time position as a professor was again at Sloan, but in the Management Information Systems group. I made a three-thousand mile detour to teach Management Science at Stanford. Later I ran the world's smallest multinational, shuttling between Boston and Europe every two or three weeks. I still spend about half my time among academics and the other half among managers. Half of each half is with technical people; the other with business people. My undergraduate degree was in English literature. My doctoral dissertation was on the psychology of intuition. My most recent book was on telecommunications.

I enjoy wandering through these different disciplines, cultures, and contexts. I learn from each of them. Throughout my career I have viewed the critical issue for the information technology field as one of synthesis and bridges. I have always believed that computers are directly relevant to the humanistic tradition of liberal thought, and vice versa. I see the artificial split between organizational and technical issues as dangerous and unnecessary, and the frequent cultural chasm between business people and information technology professionals as the one factor that can block the effective use of computers and communications.

This book is my effort to bridge those barriers. I present business design through information technology as an essential agenda for management in the coming decade, the base for a new curriculum of education for the general manager, and the opportunity to mesh what we have learned from disparate disciplines about managing and using computers and telecommunications.

I have been privileged to have many outstanding colleagues and friends in the various fields I work in. I have many "thank you's" to make to them, to say how much I value my interactions with and learning from them.

I will always owe an immense debt of gratitude to Bill McGowan, the chairman of MCI Communications. Bill provided the support for setting up the International Center for Information Technologies and helped me broaden my research, teaching, and consulting. MCI has continued to support my research, particularly in the area of the international aspects of information technology.

Roy Dingle and Ron Williams played a major role in helping shape the concepts of business integration and technology platforms that are a central theme in this book. My frequent conversations with Mark Teflian leave me in awe of the technical breadth and depth of his knowledge and creativity. Mark provided many insights on how to define the technical base for firms' IT platforms.

Jim Grant, Graham Gooding, David Horne, and Ian Scott—practicing managers in the information systems and business fields—offered me real-world situations in which to apply my ideas. Their feedback and reality tests were invaluable in ensuring that my conceptual work connected with the practical world of implementation and management.

Fernando Flores, the philosopher, software designer, master teacher, entrepreneur, and true business designer has been one of the most powerful and valuable influences on my work. I rank him the most outstanding scholar in the management field today by far. He has been generous, challenging, and always exciting.

Arthur Taylor, the dean of the Fordham Graduate School of Business, has provided me with the perfect university home and the opportunity to bring business design into the general management curriculum. My colleague Linda Jo Calloway has worked with me in presenting courses on the human side of information technology, information technology in the transnational firm, and on managing the economics of information capital.

Special thanks are due to the people who had to carry out the really tough part of getting this book into print—dealing with my handwriting, cuts-and-paste, the search for citations, and the transfer of disks. Sara Keen and Darren Hawkins are wonderfully responsive, capable, and tenacious, and probably never want to touch MS.Word again. Generally, egocentric authors like myself rage, "How can the editor massacre my prose!" I have nothing but thanks and appreciation for John Simon who structured the original manuscript and improved it immensely. My personal aide-de-everything, Anne Wells, brought an almost unbelievable degree of efficiency and organization to my work, a feat considering my ability to misplace anything, anywhere, anytime.

It is routine for an author to thank his or her spouse, often with a cutesy comment about the home life that went by the board as the scholar wrestled with writing and thinking. I thank my wife, Lynda, who was a colleague, always sharing and contributing to my book as it developed. She supplied many of the key ideas in this book, used her wide network of friends and contacts to provide me with important case sites, carried out many of the interviews reported in the early chapters, leveraged and protected my time, and gave me constant critical feedback on the many early drafts of the book. She describes our partnership as my being the thinker and she the doer. She's a top-rate thinker as well.

To you all, my warm, respectful, and appreciative thanks.

SHAPING
THE
FUTURE

Introduction

What are the new skills demanded of effective managers in the 1990s? Competence and comfort in handling information technology (IT) will be high on the list. IT—computers plus telecommunications plus workstations plus information stores—is one of the forces reshaping competition. It is an expensive resource, with many hidden costs. Those costs are growing rapidly. Investment in IT equipment grew from $55 billion to $190 billion in the 1980s, an annual growth rate of just under 15 percent. This merely continues the average growth rate in most *Fortune* 1000 firms' IT budgets throughout the 1960s and 1970s. IT continues to be the only major area of business in which investment increases substantially faster than economic growth, year after year. While IT investment expanded by almost 350 percent in the 1980s, net plant and equipment spending as a percentage of gross national product (GNP) dropped by almost 25 percent. Investments in computers and telecommunications now amount to about half of most large firms' annual capital expenditures.[1] This alone makes it part of top business managers' responsibility.

The next year or two will add even more urgency to most firms' IT planning, because it is now so integrally linked to their business planning. Well before 1993 it will be impossible for firms in the *Fortune* 1000, and for most of those in the equivalent of a *Fortune* 100,000, to define an effective business strategy that does not rely significantly on information technology. Consider just some of the almost certain business realities of the 1990s.

1) Between 25 and 80 percent of companies' cash flow is processed on line.
2) Electronic data interchange is the norm in operations.
3) Point-of-sale and electronic payments are an element in every electronic transaction processing system.
4) Image technology is an operational necessity.
5) Companies are directly linked to major suppliers and customers in electronic partnerships.
6) Reorganization is frequent, not exceptional.
7) Work is increasingly location-independent.

These are conservative, not bold, predictions.

The Business Realities of the 1990s

Between 25 and 80 percent of companies' cash flow is processed on line. Many firms have already reached the 25 percent level. Well over half of most major money center banks' revenues are derived from ATM transactions, foreign exchange trading, and electronic funds transfers. The volumes are immense. A consulting firm estimated in 1990 that $1.5 trillion of financial transactions flows through New York City's telecommunications systems each *day*. Outside of finance, airlines' passenger booking revenues flow through a computerized reservation system, more and more manufacturers and their suppliers are electronically linked to one another for all aspects of purchasing and delivery, merchants rely on on-line authorization of credit card purchases, and retailers manage inventory through on-line point-of-sale information systems linked to ordering, distribution, and delivery systems.

All these systems are part of the firms' core business activities. IT is no longer limited to accounting operations; it affects business at its very heart. Every major firm is moving rapidly toward on-line processing of its basic transactions: planning, pricing, orders, payments, engineering, design, deliveries, selling, scheduling, and so on. The 25 percent figure is a conservative estimate; most firms will be nearer the 80 percent level in cash flow processed on line.

Every time AT&T experiences a major glitch in its long-distance telecommunications network, companies find out just how much of their cash flow is on line. When a Bell Illinois switching center burned down in 1988, travel agency business dropped by 90 percent, McDonald's closed, and several insurance firms' branch offices were unable to process transactions for weeks. In January 1990, when a minor software bug prevented 50 percent of long-distance phone calls from getting through, telemarketing firms had to send their staffs home. "When it's working," said the CFO of a manufacturing firm after his company's order-entry system had been out of service for two days, "you don't know it's there, so it's easy for top management to assume it always will be. I was at fault in many ways; I was the person who blocked a lot of proposals to upgrade the network. I saw them as a cost. I now see they were really necessary business insurance."

Many senior managers today underestimate how many of their firm's basic operations depend on IT. For many, a half-day outage of the IT base would effectively bring business to a halt. Quality of business and quality of IT systems have become interdependent.

Electronic data interchange is the norm in operations. Electronic data interchange (EDI), which substitutes computer linkages for paper documents in transactions between corporations and their suppliers domestically and internationally, is rapidly becoming one of the "must do" applications of the early 1990s. At its simplest, EDI eliminates the paper chase of purchase orders, accounts receivable, accounts payable, delivery notices, and so forth (see Figure I-1).

The economics of EDI are so compelling that no firm can ignore the competitive opportunity today or avoid the growing competitive necessity within the next few years. EDI typically cuts error rates by at least half, reduces delays by days, and saves approximately anywhere from $5 to $50 per document. It has enabled RJ Reynolds to collect payments 10 days faster and to save $5–10 million per year on purchasing labor and administration. "We are hardening our stance [with suppliers]," remarked one RJR manager; "we are expecting them to be on EDI now. . . . This may be the only condition under which we will be doing business with them." In July 1990, the chairman of Sears sent letters to its 5,000 suppliers, informing them that they must adopt Sears's EDI procedures or Sears will drop them.[2]

Levi Strauss's Levilink EDI system has cut stock replenishment times from 14 days to 3 days for many of the 200,000 stores and 17,000 retailers it serves, and turnaround on delivery has been cut in half. Levi Strauss estimates Levilink has directly increased the

FIGURE I-1 Electronic Data Interchange

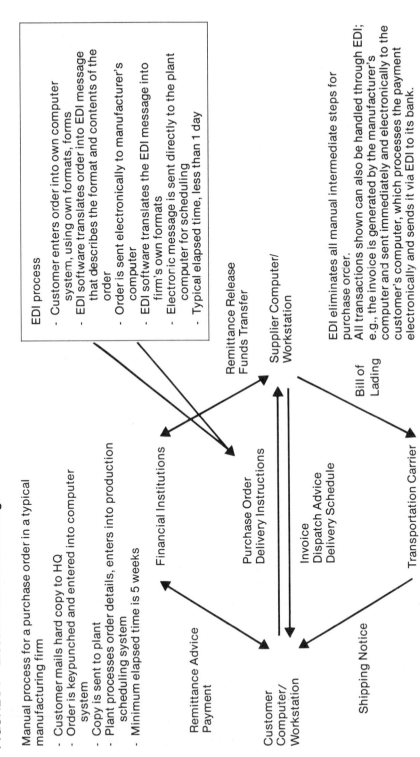

Manual process for a purchase order in a typical manufacturing firm

- Customer mails hard copy to HQ
- Order is keypunched and entered into computer system
- Copy is sent to plant
- Plant processes order details, enters into production scheduling system
- Minimum elapsed time is 5 weeks

EDI process

- Customer enters order into own computer system, using own formats, forms
- EDI software translates order into EDI message that describes the format and contents of the order
- Order is sent electronically to manufacturer's computer
- EDI software translates the EDI message into firm's own formats
- Electronic message is sent directly to the plant computer for scheduling
- Typical elapsed time, less than 1 day

EDI eliminates all manual intermediate steps for purchase order.
All transactions shown can also be handled through EDI; e.g., the invoice is generated by the manufacturer's computer and sent immediately and electronically to the customer's computer, which processes the payment electronically and sends it via EDI to its bank.

Financial Institutions

Remittance Release
Funds Transfer

Supplier Computer/
Workstation

Bill of
Lading

Purchase Order
Delivery Instructions

Invoice
Dispatch Advice
Delivery Schedule

Transportation Carrier

Remittance Advice
Payment

Customer
Computer/
Workstation

Shipping Notice

firm's sales by 5 percent. Customers benefit in many ways. One chain of stores has completely eliminated its warehouses, ordering goods in small lots as needed and having them sent directly to the store via UPS. Levi Strauss has added such features as a model stock management program that generates suggested orders for stores based on past sales and inventory patterns, plus an electronic funds transfer facility. This is just-in-time (JIT) business, using EDI to simplify and streamline the ordering and distribution cycle.

In Europe, EDI has doubled the effective speed of trucks moving goods internationally; there is less waiting at customs, fewer problems with trade documentation, and faster processing at every step in the journey. The economics of EDI are making it a business necessity. To not adopt what is a relatively simple technology is like not adopting telephones. Customers do not do business with firms that do not have telephones. Firms such as Sears have already decided not to do business with suppliers that cannot provide EDI linkages.

Point-of-sale and electronic payments are an element in every electronic transaction processing system. One leading bank lost a $400 million client in 1988 because it was unable to link to the firm's EDI systems for processing payments. As more and more manufacturing firms commit to just-in-time inventory management, orders and deliveries become increasingly smaller and more frequent. It makes no sense to manage payments weekly and by mail when the transactions they pertain to are handled second by second. It also makes no sense to add masses of expensive paperwork to manage the manyfold increase in transactions.

An airline, for instance, that manages ticket sales on line twenty-four hours a day at rates of around 1,500 transactions a second loses efficiency when travel agents' payments take weeks and involve large amounts of float and paper. They naturally start looking to "backward integrate," from their reservation system to their payments processes. Companies like Ford, RJ Reynolds, Sears, and McKesson, to name just a few, similarly extend their order-entry, distribution, and point-of-sale (POS) systems into payments; if each of these systems saves days or even weeks in business transactions, it can do the same for payments and hence for improving cash flow and internal efficiency.

Every major on-line use of IT in core operations moves firms toward just-in-time something—inventory, sales, distribution, publishing, scheduling, or reporting. Reducing time and inventory is one of the new business imperatives. Just-in-time payment is sure to be a direct follow-on to JIT inventory, very likely well before the mid-1990s.

Already, foreign exchange and electronic cash management, lock boxes, and funds transfers are core features of the corporate treasury and finance functions. Oil companies, retailers, and banks are experimenting with point-of-sale debit payments, in which customer transactions trigger direct transfer of funds. With each electronic transaction costing under a dollar, versus many times that amount for paper billing and payment by check and mail, and with the cost of capital high and operating margins ever thinner, electronic money management can be expected to be added to just about every electronic transaction. Inflation or interest rate increases in the 1990s will make just-in-time payments—moving money with transactions and coordinating funds in real time—an even more urgent aspect of business operations. The profit-margin crunch and business slowdown that is already apparent today will similarly push firms to look for ways to manage their money inventory as tightly as they manage goods inventories.

Image technology is an operational necessity. Documents are the organizational enemy. They are expensive, create bureaucracies and administrative fiefdoms, take up storage space, and add staff. Image technology reads, stores, moves, and displays any form of document electronically. It is the first widely available information technology resource that allows any of a company's documents to be managed electronically. It can, in effect, put all of a firm's relevant files onto the desks of all who need them, when they need them; there is no need to search filing cabinets or request copies from another department. The inability of transaction processing, word processing, and electronic mail to do this may be one of the main explanations of the often disappointing impacts of information technology investments on staff productivity. To date, computers have had little impact on reducing paper; indeed, they often add to it. In addition, only about 5 percent of a typical firm's documents are in "digital" form, meaning that they can be stored and accessed electronically. Image technology can handle handwriting, photographs, signatures, and even tape-recorded voice annotations to documents (such as an insurance adjuster's recorded report that is attached to a digital copy of a photograph of a damaged car).

USAA, a major insurance firm, has dramatically reduced costs and improved customer service, staff productivity, and efficiency through pioneering uses of image technology. One simple measure of benefit to USAA is an increase in policies handled per employee by a factor of five. American Express has used image technology to cut costs in several areas of customer billing by 25 percent. Northwest Airlines got a three-month payback from its investment in image processing to handle the tedious and expensive process of analyzing airplane tickets to identify payments to and from other airlines for "interlining." Continental Insurance Co. cites as just one example of improved efficiency an underwriter who wanted to see all the claims of over $50,000 filed against a certain client three years ago. What would have taken four clerks a month to find took just one person eight to nine minutes.

The potential of image technology is dramatically illustrated by the storage capacity of compact disk read-only memory (CD-ROM). An F-18 fighter plane weighs 13 tons. The manuals that come with it weigh about 13 tons as well. The manuals one company provides on CD-ROM for a commercial plane weigh about 8 ounces. A single CD-ROM disk the same size as a CD disk holds the equivalent of several hundred thousand printed pages, each of which can be accessed in a few seconds.

Image will soon be an operational necessity, in the same way that telephones are today. The 1980s were the decade of the personal computer. There is every indication that the 1990s will be the decade of image technology. The examples below will be typical, not exceptional:

> All her working life, Maureen Phillip's world has been bounded by paper. . . . Now, the paper is largely gone, swept away by a computerized imaging system, and even the staunchest computerphobes on her staff have welcomed the revolution. "Once they'd seen it in action . . . they'd go, 'God, life is so much easier.' "

> Likewise, said Gail Barnet, office services manager at the Chicago office of Continental Insurance Co.: "People are a heck of a lot happier. . . . [and] morale has improved immensely since an imaging system was installed." Of the clerical workers at pre-imaging Continental, Barnet said, "I can't see how they ever came to work every day."[3]

Companies are directly linked to major suppliers and customers in electronic partnerships. More and more companies are handling intercompany transactions—ordering, distribution, service, payments, and so forth—electronically. Suppliers link to customers, banks to retailers, multinationals to banks, airlines to travel agents, manufacturers to distributors, and hotels to airlines. This trend is hardly likely to be reversed in the coming decade. Soon there will not be a single large firm that is not electronically linked to others to the degree that daily operations depend on both parties' IT bases.

This means that companies will increasingly choose their partners on the basis of the quality and reliability of the partners' IT facilities and the ease with which the two companies can link their systems. Until now, companies have generally built their IT facilities to meet their own needs. They could choose whatever hardware, software, telecommunications and technical standards they wanted. Now they have to take into account the characteristics of their potential electronic partners' systems. Their own IT infrastructure may be a selling point or a liability.

Reorganization is frequent, not exceptional. Can any company in the *Fortune* 1000 be sure that it will not either relocate, reorganize, acquire or be acquired, and merge or divest operations? Obviously not. When any of these occur, work, and needed information, are distributed across the organizational landscape. Changes in operations require that information, communications, and processing systems adapt quickly; hence, the flexibility and scope of a firm's IT platform will significantly affect how quickly and easily it can implement a reorganization.

There are several examples of this in mergers in the airline industry. When British Airways acquired British Caledonian, it was able to mesh the two airlines' operations over a weekend because they both used the same IT architecture and standards. By contrast, the mergers of Continental and People Express and of Republic and Northwest Airlines brought chaos; the IT systems were so different that they blocked the intended integration of work systems.

Merging companies means merging operations. In a world of on-line operations, that means meshing information technology systems. No IT architecture can provide total flexibility to meet every possible combination of reorganization and mergers, but the degree of flexibility in the IT base will increasingly determine the flexibility of the organizational base.

Work is increasingly location-independent. When 25–80 percent of a firm's cash flow is on line, many aspects of work will involve computer workstations and telephones. Company personnel can just as easily be in Maine or the Caribbean as in the New York office, and the computers that store the information might be in Denver or Fort Lauderdale. Firms are recognizing the opportunity to bring the work to the people instead of the people to the work, and to locate operations on the basis of labor force cost and quality and ease of coordination. Here, information technology can provide a new source of organizational advantage that can translate into competitive advantage.

All these business realities of the 1990s represent a management challenge. They are a competitive opportunity for firms that move fast and effectively and a competitive necessity for those that delay. None of them will be easy to implement; they involve complex organizational change, careful choice of business priorities and timing, and complex technical decisions.

A Management Ill-Prepared

Senior business executives lack a well-established management process for taking charge of IT. Information technology, however pervasive, is a relatively recent arrival on the managerial landscape. Ten years ago computers and telecommunications were largely confined to back-office operations and were managed as overhead. The Information Systems department was responsible for most aspects of IT planning. Business managers consequently have not developed the kind of experience and expertise in IT that they have in finance, human resources, and accounting. The comments of senior managers at a bank that is at the forefront in using IT makes one wonder what it must be like among the laggards:

> We've still got a long way to go up the technology learning curve. I don't know much about technology but I know as much as any of [my colleagues]. . . . Top management needs a good level of understanding. It's not a choice; it's a necessity.

> I know I need the technology, but does it scare me as an old traditional banker? . . . You bet it scares me! I don't know how to measure the value for money spent on technology. When you are under pressure to cut budgets, the easiest way to do it is to cut the technology expenditure.

> Most of our top management team really don't have a clue what to do about IT. They are at the mercy of the techies. They just nod their head and hope they don't show their ignorance.

Can firms afford such ignorance of IT on the part of senior management? The IT base needed to meet the business demands of the 1990s has to be built and maintained. Decisions made today concerning information technology affect business options three, five, even ten years from now. Not being able to match a competitor's product initiatives in core areas of service because the needed technology platform is not in place has pushed a number of otherwise successful firms into a position of sustained competitive disadvantage.[4]

Interviews with executives in more than two dozen U.S. and European firms suggest that there is no lack of awareness that IT is reshaping the nature of competition.[5] These individuals readily see urgently needed applications of IT: new dealer systems in the automotive industry; enhanced point of sale in retailing; a new generation of manufacturing systems; electronic data interchange in consumer goods, insurance, and banking; and upgraded distribution links in most firms. Yet many are frustrated with IT, complaining about costs, the time it takes to obtain results, and the continuing gap between promised benefits and visible contribution to the bottom line. Few feel that they play a strong role in directing the deployment of IT; many more wish they knew how to do so.

These executives and senior managers are puzzled by IT; uncomfortable about their lack of understanding, not so much of the technology, but of the key decisions needed to exploit it. Some do not see IT as their responsibility. "I'm in the business of selling fast foods, not in the business of running computer systems," remarked one. "The less I have to deal with IT the better."

The relationship between these individuals and heads of Information Services is often one of guarded trust. Dialogue tends to be limited; in fact, communication often takes the form of monologues. Senior executives consistently demand more attention to and explanation of the economics of IT: better control of costs, more evidence of business value, and hedging of risks. Information Services managers lament that senior executives do not listen to their concerns about lead times for building key IT infrastructures that will enable business initiatives, the importance of an architecture for IT planning, and the problems of integrating separate IT resources.

Even in well-known organizations admired for their effective use of information technology as a major element in competitive positioning, the question increasingly being asked is:

What is the role of *business* managers in directing and overseeing the application of IT and building an effective dialogue with the firm's technical managers and specialists?

Few have good answers to this question, and that constitutes a huge problem. IT has become an important aspect of everyday business. It is potentially a key element in competitive positioning. But it is expensive and risky—a source of competitive fiasco as often as competitive advantage. The problem is that the typical business manager simply has no solid basis on which to decide key elements of the firm's or business unit's IT strategy.

Too often, business managers who are uncomfortable with IT try to keep the brakes on IT costs without looking ahead and asking basic questions concerning the links between business and IT planning. When do today's competitive opportunities become competitive necessities? What technology platform is needed to position the firm to handle both opportunities and necessities?

Managing the Top Line, Not the Bottom Line

Much of business managers' discomfort with IT reflects the lack of a clear economic framework for judging IT investment options and payback. Even when the competitive benefits of IT are apparent, it has proven nearly impossible to measure the economic benefits. Meanwhile, the costs of IT escalate. Many senior managers feel caught in a trap. They feel that their firm cannot afford not to invest in IT, for reasons of competitive necessity; but they also think that they cannot afford to invest without clearer evidence of its impact on financial performance.

Information technology must pay its way. It will do so if it is targeted at opportunities to improve the firm's profit structures—its "top line." Profit is traditionally referred to as the bottom line, because that is where it appears on the profit and loss statement. The implicit business model is revenue-focused. If firms grow their revenue base and keep their costs under control, they will see a healthy bottom line. Profit now has to become a top-line consideration. With operating margins being remorselessly driven down by global competition, deregulation, and sophisticated consumers, growth is becoming a way to go broke. Firms must drastically change their cost structure and rethink the relationship between revenues and profits. They need to be able to monitor sales, costs, and margins on a day-by-day basis, so that they can be alerted to and respond to trends as they occur. They need new sources of revenues that provide healthy operating margins.

There are many applications of IT that can contribute strongly and sometimes uniquely to improving the top line in these ways. The need is clear. There is no major industry today where margins can be expected to increase in the coming decade. Deregulation cuts margins per unit of revenue by up to 30 percent. This has been the case in the U.S. telecommunications and airlines industries. In Europe, the same impact on margins has been in evidence in the liberalization of the telecommunications switching-equipment market and in the far more extensive removal of restrictions in securities. Banking margins in countries like Great Britain, the Netherlands, and Belgium have dropped by about the same proportion as a result of deregulation.

Globalization also cuts margins, opening up markets to the most cost-effective firm. This has been apparent in the computer, retailing, electronic, and manufacturing industries, among others. The computer industry has seen the high prices it has been able to command for its innovations disappear down a silicon black hole; this year's software or hardware innovation is the focus of a price war next year, a commodity the year after that. As soon as a software innovation is converted into hardware and manufactured on a chip, its price plummets. Much the same thing occurs in the telecommunications industry; as fiber optics replaces copper cable, multiplying transmission speeds by factors of thousands, fierce competition drives prices down. Only in regulated environments can foreign providers use improved technology to increase margins. The gap between U.S. costs and their own then pushes businesses and consumers to demand liberalization and prices subsequently fall. Because many of these organizations are highly overstaffed, margins fall as well.

Even the top end of the pharmaceuticals industry, historically able to command high margins on the basis of its massive R&D investment, is feeling these downward pressures. In country after country, national health care policymakers are forcing either price cuts for brand-name drugs or the use of generics. Development of an effective AIDS drug would once have provided the company that brought it to market with as high a margin as it chose. No longer.

In this context, the tidy assumptions that underlie the tradition of bottom-line profit no longer hold. Unless firms' costs structures change dramatically, revenue growth will not create profit growth. To illustrate the difference between profit as bottom line and top line, Figure I-2 divides costs into three categories: traditional costs, quality cost premium, and service cost premium.

Traditional costs are the obvious costs that all firms have managed throughout this century: labor, raw materials, real estate, and so

FIGURE I-2 **Profit as the Top Line, Not the Bottom Line**

Traditional Revenue-Centered View		1990s Reality-Centered View	
	Revenues		
	↓		
	Traditional Costs		Profits
	+		↓
Less:	Quality Premium	Plus:	Traditional Costs
	+		+
	Service Premium		Quality Premium
	↓		+
Equals:	Operating Profit		Service Premium
	↓		↓
Less:	Extras, Adjustments, and Taxes	Equals:	Extras, etc.
	↓		↓
		Equals:	Profit Generator Base
Equals:	Profits		

forth. Quality cost premium refers to the often massive efforts firms have had to make to shift from viewing quality in European terms to viewing it in Japanese terms. Quality, from the European perspective, is available at a price; from the Japanese one, it is a basic requirement. The European perspective is evident in many British goods. Much of British manufacturing and services is notoriously poor, but Savile Row suits, the Connaught Grill, and all the places that James Bond shopped at provide outstanding quality, albeit at a hefty price. The European perspective treats quality as an independent characteristic of goods.

Firms can no longer treat quality as an add-on. They must pay a premium to improve it or see their position erode in the marketplace. Companies such as Ford, Xerox, and Motorola have elected to pay the premium as they have fought back against Japanese competitors.

The same is true for service. The days of customer service departments that were really complaint departments, and of banks and telephone companies that offer service with a scowl, are gone. The word "service" has even entered the vocabulary of government agencies.

How can firms cope with falling margins and eroding prices? The pricing and profit strategies employed by the airline industry since deregulation, termed "yield management," rely heavily on discounting fares and changing prices literally millions of times a day. The aim is to ensure that (1) when a plane takes off it carries the highest profit, (2) that there are no empty seats that could have been sold at another discount, and (3) no seats are filled by passengers who paid a low fare while full-fare travelers were turned away. These pricing strategies focus on profits—the top line. The airlines with the most effective yield management systems have been able to increase profits substantially in the face of drastically reduced margins.[6]

Many commentators interpret the airlines' pricing and discounting strategies as driven by the perishable nature of their product; when a plane takes off, an unfilled seat is lost forever. But yield management provides a more general lesson for business. It is practical only because the airlines have in place, in their reservation systems, a technology base that allows them to monitor purchasing activities and update prices literally by the second. Yield management is built on the core operational IT systems through which airlines sell and distribute their product.

Other industries have moved their product distribution on line in the same way. Bank ATMs, retailers and food companies' point-of-sale systems, customer/supplier purchasing, distribution and electronic data interchange facilities, and manufacturing systems all provide on-line, real-time business information that can be used to monitor and manage pricing, promotion, and production. Companies like Frito-Lay are able to provide product managers, planners, and senior executives with information on yesterday's business today and to spot trends in a few days.

The top-line framework is the basis for an economic model of IT as a business resource. This model has four business principles that should guide IT planning and economic justification.

1) Given declining margins across industries and countries and no scenario for the 1990s that will improve them, firms must restructure their cost base. This does not mean cutting costs, but rather changing the dynamics of the relationship between revenue growth and costs. IT must focus on the core business drivers relevant to managing the top line: restructuring traditional costs, providing quality and service premiums cost-effectively, and making technology an integral component of every aspect of business strategy for improving cost structures.

2) Firms must treat quality and service as basic ingredients of their operations. The guiding principle for deploying electronic data interchange, customer/supplier links, computer-integrated manufacturing (CIM), electronic distribution, payment systems, image technology, point of sale, and other applications of IT is to add to quality and service and reduce the premium cost of providing it. If IT cannot contribute directly and substantively to quality and service, it has only a very limited claim to a firm's capital and management attention.

3) Firms must shift their planning and management processes toward the equivalent of the airline industry's yield management. As more and more aspects of basic operations are handled on line, IT becomes the basis for the new management processes needed to run a business in real time. On-line customer service means no waiting, no delays; just-in-time manufacturing means exactly that; reducing product development time and time to market has become an imperative. "Time-based competition" is today's priority. Time-based planning, time-based pricing, and time-based information are their complementary and competitive corollaries.

4) Firms will need to be able to distinguish "good" revenues from "harmful" ones. They are already doing so by downsizing, going back to basics, and rationalizing. The very concept of downsizing as a positive approach to competing is new to a business culture that since World War II has lived by the ethos of growth, and stands in stark contrast to the concept of synergy that dominated business thinking in the 1960s and 1970s. Any IT innovation that aims to provide competitive advantage must ensure "good" revenues. Claims of competitive advantage must be backed by far more careful and accurate assessment of competitive costs than has been typical of the recommendations of IT gurus, teachers, and consultants.

Knowing how to mesh business and technical planning must replace the old tradition of delegating the administration of computers and telecommunications to technical specialists. Knowing when not to invest is as important as knowing when to invest. Fortunes have been lost, as well as made, on business initiatives that relied on IT.[7]

Business Design through IT

This book looks at IT as a means to a business end. It is a tool. How effectively the tool is used depends increasingly on business managers as well as on IT managers and professionals. The aim of the book is to help managers exploit the opportunities IT creates for business design—running their business better, in terms of the top line, competitive strength, and organizational health.

Subsequent chapters of the book examine seven components of business design through IT.

1) Competitive positioning through IT.
2) Geographic positioning through IT.

3) Redesigning the organization through IT.
4) Redeploying human capital through and as a result of IT.
5) Managing the economics of information capital.
6) Positioning the IT platform.
7) Aligning business and technology.

The following brief summary highlights some of the business choices and technology consequences of each of these seven components on business design.

Competitive positioning through IT. Timing is critical in using IT to either target a potential competitive advantage or avoid being put at a competitive disadvantage. Image technology, point of sale, customer/supplier linkages, and electronic data interchange are competitive opportunities today. Will they be competitive necessities tomorrow? What are the issues that determine whether, when, and how a firm decides to take advantage of a competitive opportunity or prepare for a competitive necessity?

Lead versus follow. What are the criteria for deciding when to take the lead in using IT to achieve competitive advantage and when to wait until the technology is proven and the business risk acceptable?

Compete versus cooperate. Firms must decide when it is appropriate to move ahead alone with IT and when it is advisable to join with other firms. This is often a key issue for medium-sized and small firms that cannot afford to throw money at technology.

Business "degrees of freedom." How does a firm determine what IT platform(s) it will need to ensure that it will be able to take advantage of practical competitive opportunities? Its IT choices increasingly enable or constrain its business options.

Geographic positioning through IT. As business becomes more and more globalized and IT becomes a cornerstone for international coordination of operations, companies' international telecommunications networks and computer systems will become, in effect, their organizational structure. Linking supply, manufacturing, and distribution and coordinating dispersed business teams across multiple time zones demand first-rate international communications. Yet wide variations in technology, telecommunications policy, regulations, and cost can make it extremely difficult to extend national communications and computing resources across geographic boundaries. Geographic positioning thus involves issues related not only to business, but also to where the firm chooses to locate.

Transnational computing and communication. How does a firm ensure that it has the international IT base it needs to make practical, and support, its international business strategy and to coordinate international operations?

How cities compete through IT. With the growth of IT-dependent business functions, IT will become a key determinant of whether cities attract or lose businesses. What is the basis for locating operations in specific cities to gain advantages of cost, labor supply, communication, or coordination?

Redesigning the organization through IT. IT will be used to gain organizational advantage by simplifying and streamlining communication and coordination, supporting new modes of teamwork and collaboration, eliminating unnecessary work, and reducing dependence on old organizational structures and on placing work in fixed locations. The issues here involve people as well as structure, both those who determine it and those who are part of it.

Organizational simplicity. How can firms use IT to preserve organizational manageability in an increasingly complex business environment? How can organizational advantage be translated into competitive advantage?

Repersonalization. Where can IT be used to repersonalize management and make communication more immediate, direct, and natural?

Location independence. What is IT's potential for supporting the activities of teams and business units independent of physical location?

Redeploying human capital through and as a result of IT. IT often creates and demands immense organizational and human resource changes. It changes the nature of jobs and careers. IT planning without organizational planning is almost sure to fail. Attending to the human resource issues can reduce the pain of change.

Rethinking work and jobs. Many of the disappointments from investments in IT often reflect automating the status quo, instead of rethinking what work is needed, how it can best be done, and what should be eliminated. Leading IT practitioners and commentators increasingly emphasize the need to rethink, not automate. How should the firm work back from its business ambitions and available technology to rethink work flows and jobs?

Education to lead change. How does a firm establish an education strategy that ensures that people at all levels will possess the skills and understanding required to handle waves of IT-related organizational change? How can a firm ensure that those skills do not depreciate and that careers are enhanced and not damaged as IT continues to change the business landscape?

Skill/role analysis. What new skills will be required and new career trajectories created because of IT and how will these affect recruitment, staff development, and retention?

Managing the economics of IT information capital. IT investments must be managed as business capital, not technical overhead. There are many weaknesses in current IT management processes, especially concerning how to make a reliable business case, measure the business value of IT, and manage its hidden costs. Fifteen percent per year compounded growth rates, unproven economic payoff, frequent cost overruns, new infrastructure costs, and changing technologies give rise to a host of management concerns.

Rethinking systems development. How does a company reduce the time, expense, and risks associated with informations systems development, operations, maintenance, and use?

Making the business case. How does a firm justify IT investments that provide mainly qualitative benefits and involve significant uncertainty and risk? How can the full lifecycle costs of information systems, including hidden support and maintenance costs, be determined? How can the business value of strategic IT investments be measured in order to establish criteria for evaluating trade-offs between capital commitments to IT versus other business areas?

Information access and reuse strategies. How can firms best use the information they already collect as part of their business activities for executive information systems and for new products and business planning and monitoring?

Risk management. When 25–80 percent of a firm's cash flow is on line, so, too, is its reputation, efficiency, and profit base. When the IT system is down, so, too, is the business. Technology risk is now business risk. What are the IT-related technical, economic, organizational, and business risks and how are they identified and managed?

Positioning the IT platform. Lead times for IT are lengthy—typically seven years for major business developments that require building new technical infrastructures. Companies that lack a coherent strategy for evolving a corporate platform end up with fragmented and incompatible telecommunications networks, information resources, and transaction systems. Integrating these "islands of information" has become both a business and a technical priority.

Business "degrees of freedom" analysis. What are the business criteria for defining the reach and range of the IT platform? Reach refers to the people and locations the platform can link, and range indicates the variety of services that can be directly and automatically shared across the platform. Which services and information resources need to be cross-linked? To what extent do firms that lack coordinated international IT capabilities lock themselves out of opportunities that open up abroad? What degree of reach and range is essential to meet business and organizational demands three to seven years from now?

Vendor strategies. How does a firm select the IT vendor(s) that can most effectively provide a long-term base for its IT platform and strategies? How can a firm assess its vendors' staying power?

Technology standards. Which emerging technologies and standards are critical and should be monitored? What are the business criteria for identifying high-payoff technologies?

Aligning business and technology. The competitive, economic, managerial, organizational, and technical aspects of effective IT use are interdependent and must be kept in alignment. It makes no sense to develop a powerful competitive thrust and ignore the organizational issues that are vital to making it work. It is equally foolish to build a comprehensive technical platform without the business plans to exploit it. Here the issues involve identifying and establishing relationships.

Relationships. How can the tradition of monologues and mutual frustration between business and information services (IS) managers be ended? How can a firm ensure that management policy drives the timing of competitive moves and choice of infrastructure? How do firms build the dialogue and jointly set an agenda and create plans for aligning IS and business units, corporate and business-unit IS, and external allies and partners?

Aligning core business drivers and IT springboard initiatives. How can the evolving dialogue between business and IT leaders be used to establish priorities for investment? How can firms avoid scattershot piloting of every new technology and application trend and a scattershot choice of business investments in today's technology and applications?

Moving from dialogue to action. How do we look at business processes from the perspective of ensuring customer satisfaction and target specific uses of IT to improving them? How can we design systems that integrate the work of people and machines?

The answers to these questions provide the base for a business-led strategy for IT.

Notes

1. These estimates are derived from a wide range of sources. The figures on overall U.S. capital investment are taken from published data from the U.S. Department

of Commerce. The compounded 15 percent growth rate for IT over the past three decades comes from studies in individual companies and from research by leading IT consulting firms and researchers, e.g., D. Ludlum, "The Information Budget," *Computerworld*, vol. XXI, January 5, 1987; Paul A. Strassman, "Management Productivity as an IT Measure," in *Measuring Business Value of Information Technologies* (Washington, D.C.: ICIT Press, 1988).

A number of commentators confirm the estimate of IT amounting to 50 percent of capital investment in large firms, e.g., Richard H. Franke, "Technology Revolution and Productivity Decline: The Case of US Banks," *Technology Forecasting and Social Change*, vol. 31 (1987), pp. 143–154; Nolan, Norton & Company, *Stage by Stage* (Lexington, MA: Spring 1985); Michael J. Earl, *Management Strategies for Information Technology* (Englewood Cliffs, NJ: Prentice Hall, 1989).

Many senior managers and IS managers in large firms respond to the estimate of IT now comprising half their incremental capital investment by saying, "That can't be true. It's true in our company, but it can't be that much for the others." Managers often do not know what is being spent, partly because most firms' accounting systems do not track the capital components of IT, expensing them instead, and because many of the costs are hidden and spread over many budgets.

2. Examples of applications of and payoffs from electronic data interchange are still largely anecdotal. The term is not clearly defined and in some surveys includes just about any business message sent electronically between customers and suppliers. The examples provided here come from the U.S. and European business and IT press, particularly the *Financial Times*, which regularly covers the topic and avoids hype. The figures and examples are representative, not exceptional. That said, there are many reports of firms failing to benefit from EDI. The explanations follow the old refrain about almost any new applications of IT: failure to rethink work and processes instead of just trying to automate the status quo; problems with and uncertainties concerning standards, incompatibilities, and lack of integration among relevant systems and information stores; inattention to managing the human side of technological change; and lack of appropriate senior management policy and commitment.

3. Paul Konstadt, "The Image Process," *CIO* (March 1990), pp. 33–36. The article reports other striking benefits from image technology, including a federal government agency's reducing mispayments from 3 percent to almost zero, reduction of clerical staff in a freight company by 20 percent, and elimination of an entire claims-processing department. While these examples are obviously selective, they indicate the practical target of opportunity. *CIO* estimates that about 1,300 companies were using image processing systems at the end of 1989.

4. Well-known examples of firms that were pushed into a continued competitive disadvantage include Johnson & Johnson, which was unable to respond effectively or quickly to American Hospital Supply's legendary ASAP system; Delta Airlines, which was so locked out of the travel agency marketplace for computerized reservation systems that it eventually offered to buy half of American Airlines' leading Sabre system for $650 million (the deal was vetoed by the FCC); and Bank of

America, whose rapidly eroding position against Citibank in the early 1980s was due in large part to its underinvestment in IT. For a fuller discussion of these and other companies, see P.G.W. Keen, *Competing in Time* (Cambridge, MA: Ballinger, 1988).

5. These interviews were conducted in late 1989. They are described in Chapter 1 and provide the base for much of the analysis and many of the quotes in this book. The executives were all members of the top management team in large firms that are recognized as leaders in their industry. The firms include North American and European companies in a wide range of industries. The analysis showed no apparent differences in attitudes, opinions, or expectations from IT across either country or industry.

6. The year 1986 provided a most striking instance of how computer-based yield management techniques enabled top airlines to increase profits in a context of weakened margins. Continental introduced the Maxsaver discount fare. Average ticket prices dropped by over 40 percent. American Airlines' profit went up 40 percent. Its Sabre system is no longer a reservation system but the core for its management alerting, planning, and pricing strategies. *BusinessWeek* reported in July 1990 on how Frito-Lay has similarly turned its IT base into "probably the most powerful knowledge gathering base" in the business. It captures data from stores daily and scans it for important clues about local trends. The article cites several examples of Frito-Lay being able to spot a trend in days. "Two years ago, it might have taken Frito-Lay three months just to pinpoint the problem." (*BusinessWeek,* July 2, 1990, pp. 54–55.)

Yield management represents a new method of doing business, in terms of its focus on profit as the driving element in pricing, management information, and monitoring, its rethinking of revenue/price trade-offs, and its use of IT for fast-response planning. The airline and supermarket experience seems likely to become the general practice of business within the next few years. Most discussions of yield management to date have appeared in academic management science journals and address the complex mathematical techniques and models needed to forecast and optimize yield.

7. Examples of write-offs include Merrill Lynch's and IBM's joint venture to create INMET ($250 million), Trintex, a videotext consortium ($200 million), McGraw-Hill's and Citibank's GEMCO system for petroleum product trading ($150 million), and Federal Express's Zapmail ($150 million). All these companies have been very successful in exploiting IT competitively. The expensive failures indicate that business innovation through IT is more like R&D than manufacturing, in terms of risk. As Donald Trump has been quoted as saying, "You win some and you lose some, but you've got to be in at the meeting."

Chapter 1

Management Responsibility in the 1990s

Information technology is reshaping the basics of business. Customer service, operations, product and marketing strategies, and distribution are heavily, or sometimes even entirely, dependent on IT. The computers that support these functions can be found on the desk, on the shop floor, in the store, even in briefcases. Information technology, and its expense, have become an everyday part of business life.

All of this is obvious to managers. What is less clear is how business executives can ensure that their firms benefit from new opportunities afforded by IT and avoid its well-known, oft-repeated pitfalls: botched development projects; escalating costs with no apparent economic benefit; organizational disruption; and technical glitches. Competence and confidence in handling IT will clearly be key to effective management in the coming decade. Senior executives can no longer delegate IT policy and strategic decision making to technical professionals.

Now that IT is already the single largest element of many firms' capital expenditures and continues to grow as a percentage of their operating budget, it is senior management's job to make sure that IT pays its way and that IT investments are targeted appropriately. Their IT managers and advisers cannot be expected to set business priorities. Whether approved for reasons of competitive necessity or with the specific aim of stealing a competitive lead, IT can now strongly influence a firm's position in its industry. This warrants a new level of involvement and direction by the company's business leadership.

Senior managers have growing organizational as well as financial and competitive responsibilities. IT increasingly changes jobs, skill needs, work, and relationships. Technical change has become synonymous with organizational change. Such change can be complex, painful, and disruptive. The people side of IT is often more difficult to anticipate and manage smoothly than is the technological side. The leading IT practitioners, consultants, and researchers invariably expound on the organizational challenges and problems associated with technology opportunities.

Meanwhile, the technology continues to change at an exciting and bewildering pace. Betting on new IT innovations can mean betting the future of the company. Leading-edge firms are sometimes said to be on the "bleeding edge." Almost any business executive is aware of disastrous projects that had to be written off, often after large cost overruns, because the promised new system simply did not work. Yet firms content to comfortably trail along risk falling behind the best business practice in their industry. Some leaders in applying IT to a new business application have gained up to a seven-year window of advantage.[1] The successes of firms such as Citibank, American Express, American Airlines, Federal Express, American Hospital Supply, and McKesson in the mid-1980s largely reflected a combination of aggressive business leadership that explicitly included IT as a business enabler. All of these companies made mistakes with IT as well as creating successes. Federal Express wrote off several hundred million dollars on its Zapmail project, which aimed at establishing a global facsimile service via satellite. A "bug" in American Airlines' yield management software cost it $50 million in lost revenues. Citibank had to abandon a joint venture with McGraw-Hill to create a new petroleum-product trading system. As with R&D, IT offers no cozy guarantees. As with R&D, failure to invest in it today can mean mortgaging the competitive future.

In this context, the management process will determine how much competitive edge firms derive from IT. All firms have access to the same technology base. Smaller ones can often exploit it faster than large ones, now that personal computers and related hardware, software, and telecommunications have decentralized many aspects of IT. Senior managers who provide active policy direction become enablers for exploiting IT. Those who do not can easily become blockages. The most common complaint—and a very widespread one—among leading Information Services managers is the lack of management action at the top. They report inertia, avoidance of making decisions, and an overattention to narrow cost control.

It is natural for today's business leaders who are often skeptical about IT, having seen too many broken promises and wasted investments, to err on the side of caution. Because IT has not been a significant part of their experience as they have moved up the management ladder, they often see its costs but not its benefits. They put the brakes on what they see as runaway costs. Others may not block progress but instead passively react to the business, economic, and technical changes IT precipitates rather than actively take charge of them. In doing so, they lose the chance to use IT to shape the firm's future.

Competitive, technical, organizational, economic, and managerial choices and consequences are now so interdependent that they cannot be handled in isolation from one another. A creative competitive strategy that ignores organizational issues, or a technical strategy that ignores the complex economics of IT, is not just risky but very likely to fail.

To be effective, business design through IT must balance the interplay among these elements. The decision agenda for top management has to include consideration of competitive and geographic positioning, redesigning the organization, redeploying human capital, managing the economics of the firm's information capital, positioning the firm's technology platform, and ensuring alignment between business thinking and action and technical planning and implementation.

Senior managers need to recognize which are the truly critical and urgent decisions about IT that only they can make. They need to anticipate key IT-engendered human resource and organizational issues and establish business and organizational policies to guide technical planning. They need to evaluate technical decisions in terms of which ones will most directly influence the range of business options in the 1990s, either opening up potential opportunities or closing them down. And they need to be as effective in doing so as they are in setting the criteria for financial, marketing, human resource, research and development, and manufacturing strategies.

The business design perspective brings to the foreground issues of timing that are at the very heart of IT choices and consequences, issues that raise questions such as:

When should we take the lead in business initiatives that depend on IT? When should we follow our competitors?

How early must we begin such an initiative? How long is it likely to take before a given competitive opportunity becomes a competitive necessity?

Given the long lead times inherent in building IT infrastructures, how can we mesh the timing of IT investments with rapid and often unpredictable business and technological change?

What kind of technical infrastructure will we need in the coming years? What business assumptions, priorities, and guidelines should drive its development and how long will it take to implement? Can we anticipate organizational impacts? Can we move ahead of the change curve?

These questions have to be answered at top management levels. IT managers can act as advisers here, but these are business policy issues. Lead times are long for IT development, and in the whirlwind business environment of the 1990s, today's delayed decision will become tomorrow's default decision. In 1995 a firm will not be able to make up for decisions it did not recognize it needed to make in 1992. Managing IT lead time is already a major factor in gaining or keeping a competitive edge, and letting decisions slide by default is a corresponding competitive handicap.

The Management Process: Beyond "Awareness"

The worlds of IT professionals and business executives have occasionally crossed but rarely converged. Fewer than 10 years ago—a fraction of a senior manager's career and a very short period in terms of business and organizational change—computers were viewed as part of the firm's overhead and managed as such. Telecommunications meant telephones; personal computers were uncommon; information systems professionals' skills, interests, and careers focused on software development and project management; and business managers could avoid or ignore anything to do with computers or delegate it to the firm's Information Services function.

Figure 1-1 depicts the management process for IT planning as a learning curve with three distinct levels:

1) Awareness of the business importance of IT, in terms of its potential impact on the basics of competition;
2) A business vision to drive its deployment; and
3) A compelling business message for building a comprehensive and "integrated" IT platform rather than a set of separate facilities and services that are "incompatible."

FIGURE 1-1 Management Decision Process for IT

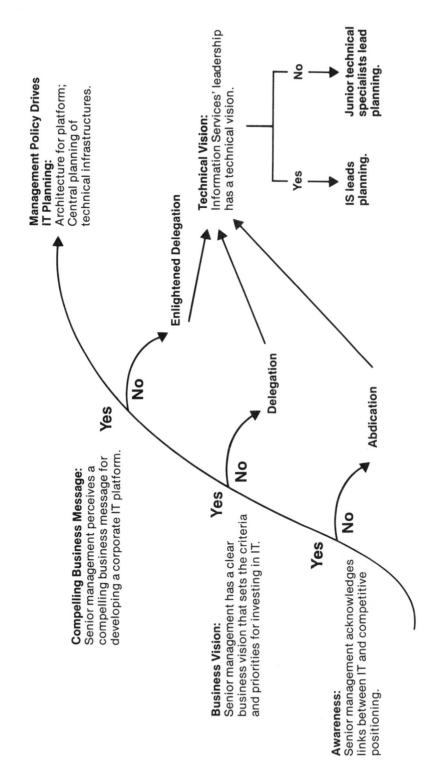

Management Policy Drives IT Planning:
Architecture for platform; Central planning of technical infrastructures.

Yes

Compelling Business Message:
Senior management perceives a compelling business message for developing a corporate IT platform.

No

Enlightened Delegation

Technical Vision:
Information Services' leadership has a technical vision.

No → **Junior technical specialists lead planning.**

Yes → **IS leads planning.**

Yes

Business Vision:
Senior management has a clear business vision that sets the criteria and priorities for investing in IT.

No

Delegation

Yes

Awareness:
Senior management acknowledges links between IT and competitive positioning.

No

Abdication

The framework was developed from interviews with over 80 top managers in 27 companies in North America and Europe in 1989. They included the chairman and/or CEO, other members of the top management team, such as the head of finance, and the head of Information Services. The companies are all recognized as among the business leaders in their industry. The interviews addressed how top managers think about IT as a business resource and their own role in the IT planning process.[2]

The interviews provided a rich picture of the perspective of business leaders on IT. They revealed more frustrations and concerns than comfort. These executives are well aware of the new business relevance of IT. They are not Luddites. Most of them, though, have no clear picture of exactly how IT can be exploited most effectively. They lack a personal business vision to drive it. They look to their IS and business-unit managers to take the lead. They adopt a short-term focus on IT, mainly because they are skeptical about budgets and schedules and see costs as nearly being out of control.

In many of the companies, there is a growing sense of mutual frustration between the head of IS and the top management team. The CEOs are puzzled by the constant problems of predicting and managing IT costs and disturbed that the large amounts of money they have committed to IT may not provide a real economic return. Their senior IS managers are just as disturbed that the company's future may be at risk because there is no corporate blueprint for building the IT infrastructures needed to compete effectively in the 1990s.

Both parties want a true dialogue about how best to manage and use IT. Most of the companies have invested in education programs and research studies to build management awareness of IT as a competitive weapon. The interviews show clearly that awareness does not necessarily lead to a business vision for deploying information technology.

The term "vision" has in recent years supplanted strategy and forecast in the management literature and planning documents, perhaps reflecting a sense among managers and observers that no one can reliably anticipate the business context of the next decade. Vision provides a broad-directive target to help firms plan when they cannot easily predict. Facing rapidly shifting business demands and technology and the need to maintain, operate, and integrate expensive existing technology investments, Information Services managers need a guiding vision. If they are not able to see what the business leaders want to accomplish and where they place priorities for innovation, they either do the best they can, often picking up conflicting signals, or try to create and communicate their own technical vision for the firm. Their basic problems are the long lead times inherent in IT and the massive technical problems involved in integrating separate IT resources and applications. The business vision is the blueprint for tackling these problems.

Such a vision is like a photograph of the future that the business leaders want to create—"This is what we will become. Here is where we will focus our efforts. These are the factors every key investment should emphasize." The vision is a set of simple directive statements about business imperatives, fundamental priorities, and targets of opportunity—a bedrock statement of purpose that, though it may not be a forecast or plan, sets the priorities for business planning and establishes the criteria for IT investments. It shifts the basis for making the business case for IT investments beyond cost displacement and routine automation and permits greater dimensions of potential competitive edge to be seriously assessed. It helps move the terms of discussion from passive to active.

A critical policy decision that is far-reaching in its competitive and organizational impacts is whether the business vision points toward a compelling business message for building a comprehensive and integrated IT platform. Frequently, such a message is lacking; business executives then encourage a piecemeal approach to IT, with business units implementing applications on a case-by-case basis. This was by far the most frequent attitude to IT planning among the business executives interviewed.

There are many sound reasons for the application-by-application approach, particularly the value of decentralizing business-unit responsibility for IT decisions, exploiting decentralized IT components, especially personal computers, and matching the specific choice of technology to the characteristics of the application. Few firms want to go back to the old days of a highly centralized data processing fortress, run by managers who were stereotypically lacking in business understanding and interest.

There is a practical alternative to the two extremes of fragmented decentralization and monolithic centralization. It requires strong top management endorsement and a long-term view of IT as a business enabler. It separates the IT infrastructure from the specific applications to be built on it. It establishes the policies for designing a corporate IT platform that can be the base for a wide range of applications and services.

Such a platform will be based on an "architecture" that imposes common organizationwide technical standards for choosing equipment, software, and telecommunications services. These should have the same force as accounting rules. An effective architecture will balance central coordination of the IT infrastructure with decentralized use. The more flexible and adaptive it is, the wider the variety of shared and interlinked resources it enables.

Awareness: A Starting Point

Changing the management process and dialogue about IT begins with development of an awareness that there is an actual or potential link between IT and key aspects of competitive positioning. Without such awareness, IT is fundamentally irrelevant to an executive's personal agenda. That was the situation in most firms through the early 1980s. Even today, many business executives pay only lip service to the business importance of IT; the evidence is the amount of time they spend on it.

Consider, for example, the CEO of a major foods company whose 1989 mission statement, "The Year of Decision," discusses the urgent need to reduce costs and improve quality of service without a single reference to IT. He instructed his newly appointed head of Information Services to "tell me if we can use IT across the company to gain a competitive edge." What does that mean coming from a CEO whose mental map does not include IT as part of "the year of decision," even though he is in an industry where leading firms are already using IT to transform distribution, customer service, and management information?

Another CEO in the petrochemical industry, though he is very interested in the use of IT in oil-field exploration, dismisses as "irrelevant" the idea that point-of-sale and pump-island automation will become key competitive differentiators in the selling of gasoline. Mobil and Shell do not seem to think so, and the experiences of banks, retailers, airlines, automotive manufacturers, and distributors over the past ten years strongly suggest that electronic point-of-convenience customer service will be one of the requirements of effective competition. But this CEO and his top management team are impatient with IT and with IT people. The company's two top corporate IS managers were pushed into early retirement. Now senior managers no longer have to listen to them, or to think much about the issues they persisted in raising.

During the 1980s, much of the popular literature on IT and the

educational efforts of leading academics and consultants emphasized building management awareness. This is a limited, though
essential, starting point. Awareness is not action. It can amount to
little more than: "That's interesting. I wonder what our IS people
are doing about it." Unless the people at the top accept that IT may
end "business as usual," there is very little an IT manager can
accomplish except by stealth.

The Business Vision: Seeing and Believing

A simply stated business vision can create an entirely different
perspective on IT. For example, the CEOs of three insurance firms
have different thoughts about their own business priorities and
what they imply for IT. The first envisions his company becoming
"the easiest in the industry to do business with." He wants customers to "see a total difference between dealing with us and with
anyone else." By emphasizing cost, speed of writing policies, and
quality and responsiveness of customer service, he hopes to be able
to sell customers all of their insurance needs.

It is hardly surprising that for more than 20 years this firm has
been a leader in applying IT in areas where other companies have
lagged: building data bases that profile customer demographics and
purchasing patterns; providing telemarketing and cross-selling over
24-hour 800 number customer service telephones; producing high-
quality customer annual reports that summarize account activity;
offering automatic credit card payment options for premiums; and
cross-linking insurance and investment products. The case for IT
capital investment is made in terms of business logic and in terms
of the technology opportunities implied in the vision.

The vision of one of the other two CEOs is entirely different and
suggests quite different directions for applying technology. This
CEO, conceding that his company has been an average performer
in most areas of insurance underwriting, selling, and service, has a
vision of the company becoming a star investment manager and
being able to offer aggressive pricing. "We will not lead the industry
in products; in fact, we will offer the same as everyone else, perhaps
a year or two after they first introduce a variant on a product or
even a new product. We will, however, offer the best price and
accumulation of value—that, in the end, is what people want. In the
insurance industry, you are basically investing today's premiums to
cover tomorrow's payouts. If you outperform the others in manag-

ing your investment portfolio you can charge less for the premiums and pay out more when policies mature."

Given this business vision, the obvious target of opportunity for IT is cash management, trading systems, expert systems for portfolio management, and analytic decision support systems. Though the company is in exactly the same industry as the other insurance firm, the different vision encourages and enables a very different exploitation of IT. In the other firm, it would be hard to make a case for an international telecommunications network to support the investment function, but easy to secure top management commitment to major expenditures on image technology that pulls customer documents together into a single electronic file, everything from letters and checks to payments and photographs. In the second firm, the decision would go the other way.

The third CEO, who heads one of the ten largest insurance companies in the United States, has no vision for IT. His stated business strategy is to "make sure the basics are not neglected: underwriting, investment, the agency distribution system, pricing, and costs." He views IT as part of the responsibility of the heads of individual lines of business, and sees no need for him to articulate a statement of direction. His attitude to IT is perhaps best described as amiable drift.

Business units in this company vary widely in IT spending, quality of systems, and degree to which IT use is a central rather than peripheral component of business innovation. There is a broad variety of systems in place, several of which push the technical state of the art and several of which are almost museum pieces. None of the major business units' key systems can directly share information with the others. They use different hardware and software and at last count had over 30 separate telecommunications networks in use. In the jargon of the IT field, the networks are "incompatible" instead of "integrated."

It is dawning on most of the senior managers of these business units that the company is in serious trouble. It lacks the coordinated IT planning and platform needed to cross-sell individual products, reduce transaction costs, meet corporate customers' demands for more efficient and timely service, speed up the issuing of policies, upgrade the agency system, and solve the many recognized problems of bureaucracy, overstaffing, and inordinate cost growth. There is plenty of IT in this firm, but no business vision to make it more than an ad hoc and fragmented tactical resource.

Recently the company hired as head of Information Services one of the rising stars of the IT field. He immediately declared a state of emergency; he purged the IS organization, canceled a number of major development projects, and announced that unless the firm developed a corporate architecture for IT it should forget about competitive advantage for at least the next five years. No business vision means no real opportunity for IT to provide competitive value. The new head of IS is realistic on this point:

> I cannot create the business strategy. If it's there, I can make sure we in IS rev up our engines and add value. I believe I was brought in only because [the CEO] is getting messages from the organization that something is missing and he doesn't know what it is. I don't see him as the person who will make things happen. He may prevent them from happening. He may give them his blessing. . . . It's a real mess around here. It's not the IS folks who made it, but they are getting the blame.

The Compelling Business Message: Going All the Way

Unless the business vision delivers a compelling business message for a shared corporate IT platform, there is every reason to allow individual business functions, operating units, subsidiaries, and national operations to make their own choices about IT. Insofar as a decentralized business philosophy demands decentralized IT planning and implementation, this makes business and organizational sense. Decentralized technology—personal computers, workstations, departmental systems, local area networks, and so forth—supports decentralized operations.

This argument, however attractive, is deceptive to the extent that it overlooks business trends of the past decade that are increasingly dependent on information sharing and shared communication systems—for example, trends such as: the shift from largely independent business functions to interdependent ones and from product- to relationship-based services and cross-selling; widespread emphasis on teams and cross-functional cooperation; and customer expectations that information will move as fast as goods and transactions. To share information across products, services, locations, companies, and countries requires a shared platform and an end to separate IT applications for separate business activities. Regardless of the many valid business and technical reasons for separate technology bases for particular units, services, and types of applications, there are compelling reasons for interlinking them in a shared corporate platform that makes information a real business resource.

The lack of a corporate IT platform is today blocking realization of the business vision of one of the world's top manufacturers. Each of the company's four main IT facilities—engineering, production, dealers, and finance—is based on entirely different technology and vendor architectures. None can share information directly or link across the firm's locations and customer base. The company is building a new dealer system that must be able to "backward integrate" into production's inventory and scheduling systems, finance's credit and payments systems, and engineering's quality control systems. It cannot do so, any more than a Betamax tape can be played on a VHS VCR. Top management is upset and is asking how the firm got into this situation. The answer is that it has plenty of business vision but no compelling message for a shared IT platform that would provide the infrastructure both for integrating existing applications and for developing new services made practical by its architecture.

Most leading Information Services managers think in terms of "architecture" and "integration." The interviews referred to earlier show that few of their bosses do. They imply that only about a third of *Fortune* 1000 firms have a platform-centered strategy for IT. In the sample of 27 firms, exactly a third (9 firms) had a business vision to drive IT. In the other 18, responsibility was delegated to IS and business units.

Delegation guarantees that there will not be a platform. The contrast between delegation and vision is marked, both in its implications for the resulting management process and in the choice of IT strategy and technical base. Listen to a couple of CEOs who have heard a compelling business message and a couple who have not.

> If you have an architecture and the others don't, then you create a source of competitive edge for you and a problem for them. (CEO of a major distribution firm)

> We are positioning to be the provider of travel-related services. That's impossible unless you have the platform. That's what [the firm's augmented computerized reservation system] is—our platform for doing business. (CEO of a leading international airline)

> To be honest, I am not at all familiar with the details of our IT planning. . . . I expect [the head of Information Systems] and the heads

of the U.S. functions to manage the strategy. (CEO of another international airline)

I see no need for an IT architecture. That is nothing more than another technology bureaucracy. There is no need for someone at my level to spend time on IT. The divisions must make their own plans. (CEO of a consumer goods company)

Would any top manager openly say, "To be honest, I am not at all familiar with the details of our finance planning," or "There is no reason for someone at my level to spend time on marketing"?

Management Awareness and Technical Planning

The management learning curve depicted in Figure 1-1 moves from a passive to an active view of IT and from delegation to taking charge of IT at the policy level. Management awareness drives progress up the curve. Senior management that lacks awareness of an actual, likely, or potential linkage between the firm's business positioning and its IT choices tends to abdicate responsibility for IT decisions to the head of the Information Services function (or his or her technical specialists) or, in a decentralized organization, IS staff and business-unit managers. Under conditions of abdication, decisions and recommendations will almost certainly be made on the basis of short-term budgets, tight cost restraints, and, increasingly, efforts to cut back what is seen as an area whose costs persistently increase faster than can be justified by business growth.

One of the most cherished assumptions of the proponents of IT as a major source of competitive advantage, and the logic for many articles and management awareness education programs on the topic, has been that increasing senior management awareness would increase and improve management action.

Figure 1-1 suggests that this assumption is not necessarily valid. Awareness may just lead to delegation. Senior managers who see IT as important to the firm but do not believe they must play an active role in policy and planning push responsibility down. Lacking a directing business vision to drive IT, but possessing awareness that makes them receptive to new ideas and proposals, they provide, if not a green light, at least a cautious yellow. IS management then makes basic recommendations and de facto decisions about IT under less constraining circumstances than apply under conditions of top management abdication.

Awareness without a guiding vision encourages proposals for specific IT applications on a bottom-up basis. Engineering may propose initiatives in computer-aided design, finance in electronic cash management, and sales in telemarketing. IS and business-unit leadership generally work closely to develop IT applications that support the business, often in important competitive domains. Engineering may use computer-aided design (CAD) systems to shorten development cycles and improve quality; sales may use telemarketing as a basis for aggressive targeted marketing programs; finance may implement a worldwide cash management and reporting system. Each application is essentially separate from the others, and will probably be incompatible with them since the vendors best able to satisfy engineering's needs are not likely to be the best vendors for finance.

Though such an approach may be cost efficient and reduce technical risk and political strain, it seldom creates a sustainable competitive edge that significantly affects a firm's companywide strength and bottom-line performance. If the system took two years to build, it will take competitors no more than that. If it used a purchased software package, it will be all the more easily imitated.

Such investments are not positioned to be leveraged through cross-linkages or a multi-use delivery base. The IT strategy is a set of competitive applications rather than the development of a platform on which a stream of competitive applications can be built, combined, interlinked, and augmented.

An aggressive application-by-application approach results in a rigid set of stand-alone systems, each competitively valuable in a given context at a given time. When the time and context change, the systems can become a barrier to adaptation, especially if top management has encouraged decentralization of IT planning and implementation to support a decentralized business environment.

Consider the U.S. bank that split its corporate IS department into 10 divisional IS units to match its 10 strategic business units. This approach matched the unit IT resource to the unit business need, which in this company meant matching it to the current organization and product groupings. The bank is now finding it difficult to change its organizational and product strategy. Efforts to shift toward cross-selling and new forms of relationship management, to move to a different decentralized structure, and to create new products by combining information from different separate systems are being stalled by the prohibitive cost of adapting the separate IT bases, or even scrapping parts of them and rebuilding.

IT service delivery bases are as hard to change as office building complexes. Once specific equipment has been chosen, specific software built, specific data bases defined and created, and specific telecommunications networks adopted, incompatibility becomes the norm, not the exception, and restructuring and sharing become a bigger and bigger problem.

The growing emphasis among IS managers on IT standards, architecture, and integration reflects a recognition that firms cannot continue to add incompatibility and complexity to their IT bases. "Systems integration" is a new trade, with its contractors and consultants, many of them among the outstanding technical planners and implementers in the field, earning large fees to fix the mess of incompatible technical systems that block business priorities. There is a strong move across the profession to foster technical standards leading to "open systems" that will ensure full interoperability and interconnection of equipment and applications. These standards will not be defined and implemented easily or quickly, and there is widespread uncertainty and controversy over many of the details of the standards, vendor capabilities and commitments, and the promises and risks of integration. That said, for the leading IS managers and their advisers, integration is their main priority for the 1990s.

The task of imposing a coherent structure and creating a practical path to integration generally begins with analysis of the longer-term corporate business vision and direction, analysis that has been largely lacking in many firms. The recent and rapid emergence of systems integrators reflects a management failure; they should not have been needed. The long-term costs of ignoring architecture and integration have been obvious for at least the past ten years. Many IS heads have tried and failed to get the message across to senior executives. Abdication and delegation result in managers looking only at the next application, not the overall framework for applications.

They also generally leave IS to make what amounts to business decisions. Without a guiding senior management vision and commitment to an integrated platform, abdication and delegation shift the locus of decision making about the choice of an IT base to the technical planners. Business-unit managers may establish their own requirements and be in a position to fund and implement them, but it is IS, not the business leadership, that drives the planning process. The dialogue will generally be limited to IS proposing and business management signing off. IS then has to shape a technical vision. This is not an easy task, even on an application-by-application or business-unit-by-business-unit basis.

"Information technology" is a relatively new term. Until recently, the many components of IT were separate technologies (e.g., computers, voice communications, office automation, personal computing, data communications) and largely separate organizational responsibilities, skill sets, and experience bases. It is impossible here to give even a flavor of the current problems involved in fitting these technologies together. Yet a major IT facility that hopes to achieve competitive advantage, or to forestall competitive disadvantage, has to combine most or all of them. The many relevant complexities, vendor-specific strengths and weaknesses, cost/performance trade-offs, and uncertainties implicit in combining different technologies demand a technical blueprint based on technical vision.

Does the head of IS have a technical vision? If senior management, lacking a guiding business vision, has abdicated or delegated decision making to an IS leader who does not have a clear technical vision, key decisions will be made by the firm's technical specialists. Absurd as this may seem, it is a natural consequence of senior management failing to provide the driving business vision and the IS manager being unable to keep current in a fast-moving and ever-changing field. To have reached this circumstance is for the firm to have turned the business decision about IT and the IS decision about the technology platform to support the business into a technocentric and analytic review, in which the opinion of the most knowledgeable and technically up-to-date (and hence, often young) specialist determines commitments that can shape or constrain basic aspects of the firm's business future. It happens often.

In the foods company, mentioned earlier, the CEO wanted to know if and where IT could create a source of competitive advantage for the firm. The new head of IS focused on establishing relationships with the heads of the key business units. He was not knowledgeable about many aspects of IT developments and relied on the advice of his head of systems development. She had very firm and well-thought-out opinions on choice of standards and of hardware and software vendors. She took a strong position, favoring a direction that relied on unproven technology and very specialized development tools. She had almost no business experience and even less interest in business. She was a technology bigot.

Her boss went along with her recommendations. They may or may not turn out to be sound, but the absence of a business vision for IT plus no technology vision means that this technical specialist may have defined the distribution, communications, electronic data interchange, and point-of-sale strategies of a $3 billion firm for the next ten years.

The Dimensions of the IT Platform: Reach and Range

The concept of an IT platform is one of the central components of business design through information technology. How to define and build the platform is described in detail in Chapter 7. The key features are summarized below.

1) The firm's IT platform is a major determinant of its "business degrees of freedom" for the 1990s; the business functionality of the platform will determine which IT-dependent products and services are practical and which are not. The platform enables or disables future business options.

2) The platform is a shared information services delivery base, whose business functionality is defined in terms of the two dimensions of reach and range (see Figure 1-2).

3) Reach determines the locations the platform can link, from local workstations and computers within the same department to customers and suppliers domestically, to international locations, or—what is an impractical ideal today but perhaps not five years from now—to anyone, anywhere.

4) Range determines the information that can be directly and automatically shared across systems and services. At one extreme, only systems built on exactly the same hardware and software can process messages and data created by each of them. At the other, any computer-generated transaction, document, message, and even telephone message can be used in any other system, regardless of its technical base.

The more critical information technology is as an element in any firm's opportunities or necessities in the areas of competitive positioning, geographic positioning, organizational redesign, and human capital redeployment, the more dependent it will be on the degree of reach and range in its IT platform.

Is IT a key factor for the firm's future business and organizational health? Can IT help manage profit as the top line and not the bottom line, contribute to restructuring costs, and guarantee quality and

FIGURE 1-2   **The Business Functionality of the IT Platform:
Reach and Range**

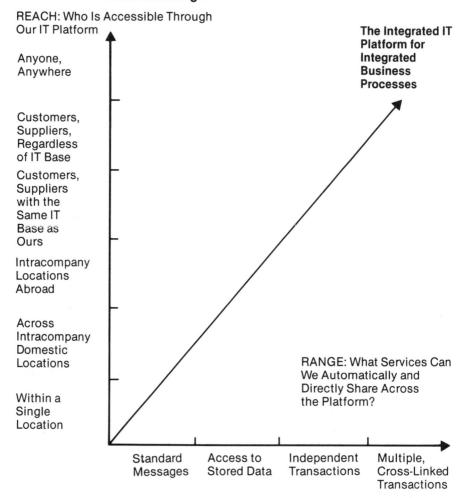

REACH: Who Is Accessible Through
Our IT Platform

**The Integrated IT
Platform for
Integrated
Business
Processes**

Anyone,
Anywhere

Customers,
Suppliers,
Regardless
of IT Base

Customers,
Suppliers
with the
Same IT
Base as
Ours

Intracompany
Locations
Abroad

Across
Intracompany
Domestic
Locations

RANGE: What Services Can
We Automatically and
Directly Share Across
the Platform?

Within a
Single
Location

Standard   Access to   Independent   Multiple,
Messages   Stored Data   Transactions   Cross-Linked
                                                     Transactions

service? Does the firm need a platform? Should top management make it a priority? The aim of this book is to help business executives answer these questions. The questions can no longer be avoided.

Notes

1. For a fuller discussion of the nature and duration of competitive advantages, see C. Wiseman, *Strategy and Computers* (Homewood, IL: Dow Jones-Irwin, 1985);

P.G.W. Keen, *Competing in Time* (Cambridge, MA: Ballinger, 1988); M.J. Earl, ed., *Information Management: The Strategic Dimension* (Oxford: Clarendon Press, 1988).

2. The composition of the companies and managers interviewed is shown below.

	Companies in North America	Europe	Most senior executive interviewed
1. Automotive	X	X	Chief Financial Officer
2. Insurance		X	Chairman
3. Airline		X	Chairman
4. Petrochemical	X	X	CEO
5. Pharmaceuticals		X	Chairman
6. Bank	X		CEO
7. Telecommunications	X		Chairman
8. Distribution		X	CEO
9. Retailing		X	CEO
10. Distribution		X	CEO
11. Insurance	X	X	Chairman
12. Pharmaceuticals		X	CEO
13. Airline	X	X	Chairman
14. Market Research		X	Chairman
15. Petrochemical	X	X	CFO
16. Bank	X		CEO
17. Bank	X	X	Chairman
18. Insurance	X		Chairman
19. Manufacturing	X		CEO
20. Airline	X		CEO
21. Publisher	X		CFO
22. Pharmaceuticals		X	Chairman
23. Bank	X	X	CEO
24. Consulting	X		CEO
25. Insurance	X		Chairman
26. Realty	X		CEO
27. Bank	X		CEO

Breakdown by industry:

Banking	4
Distribution	2
Insurance	4
Manufacturing	2
Petrochemicals	2
Pharmaceuticals	3
Publishing	2
Retailing	1
Other	4

The interviews reveal no consistent differences between industries or patterns within them. They also show no marked differences between U.S. and European companies.

Chapter 2

Competitive Positioning through Information Technology

Much of the literature on information technology in recent years has focused on the search for competitive advantage. Indeed, "IT-and-competitive-advantage" became almost a breathless single word in the 1980s. An entire industry of advisers and writers grew up around it. Mainly they offered guiding frameworks for identifying competitive opportunities. The simple table in Figure 2-1 captures the essence of the frameworks.

The labels on the axes vary, but the logic is much the same: gaining competitive advantage rests on choosing the IT application that provides the greatest likely strategic benefit. That should be common sense—why on earth work on one that provides the least potential? Surely, no one would consciously target the cells that reflect low but safe business payoff and impossible-to-deliver high payoff.

In practice, most firms did just that for much of the 1970s and 1980s. History and technology, not business logic, drove the deployment of IT in what might be termed the overhead era, when most applications of IT addressed clerical operations. IT budgeting and accounting were handled on an annual basis, and the head of IS was clearly a mid-level staff position. The IT function was characterized by detachment from senior management thinking about business direction and priorities. IS managers, not business executives, wrestled with the problem of how to identify high-

FIGURE 2-1   **Generic Frameworks for Identifying High-Payoff IT
Opportunities**

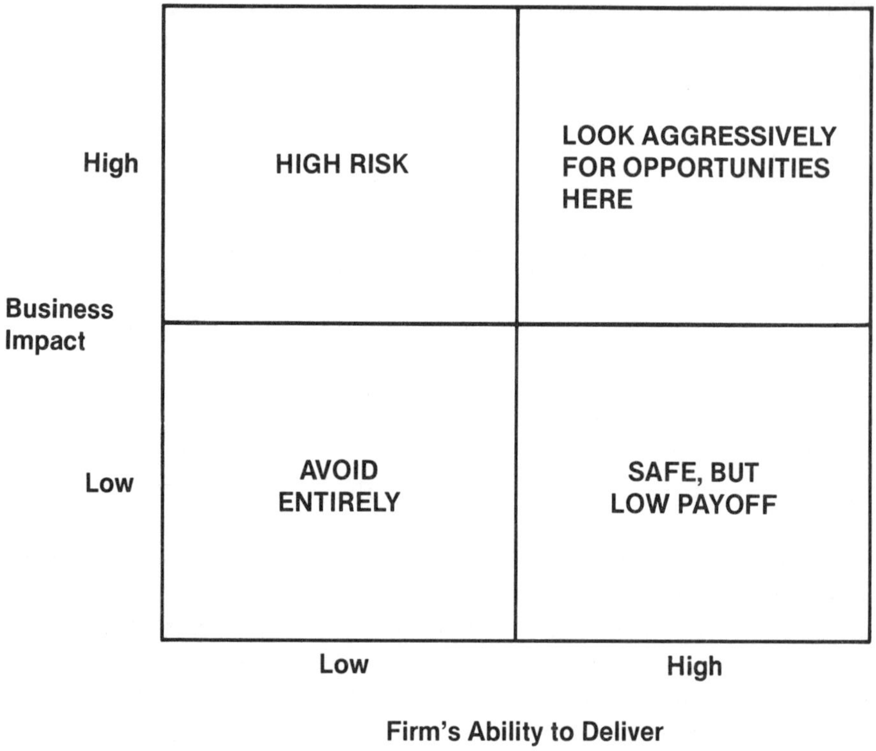

	Low	High
High	HIGH RISK	LOOK AGGRESSIVELY FOR OPPORTUNITIES HERE
Low	AVOID ENTIRELY	SAFE, BUT LOW PAYOFF

Business Impact (vertical axis)

Low   High

Firm's Ability to Deliver

payoff business opportunities for applying IT. They also struggled—and still do—with the difficulties of software development;
for the 1970s, no large system fell into the category of "easy to
deliver."

This history is reflected in the emphasis on large-scale automation
of clerical operations that dominated data processing during the
1960s and 1970s. Computers were expensive and the business case
was made in terms of improving costs and efficiency in activities
that involved a heavy volume of transactions. "Data processing"
was construed as applying automation to assembly-line clerical activities that began with the collection of transaction documents, required routine calculations, and generated printed documents.
From this perspective, payroll was a natural starting point for data
processing, and accounts receivable a logical next step.

Over time, this initial choice of applications locked most information systems units into (1) a heavy maintenance load, (2) projects that built on past investment, and (3) a human resource base that was comfortable with the status quo. Payroll and accounting systems maintenance, one of the most resource-consuming tasks many IS groups perform, will have to continue, if not for eternity, at least until there are no more changes in federal and state income tax rules, employee benefit plans, salary scales, and union contracts.

During the overhead era there was little opportunity or incentive for IS or business units to invest outside their sphere of experience or political control. The nature of the available technology and its cost limited applications to either number crunching or data processing. Most IT uses thus fell into the zone of limited business impact/high IS ability to deliver (the bottom right-hand corner of Figure 2-1).

In the late 1970s, for example, most large firms had well-established sales reporting systems, but few had systems for competitive monitoring and alerting. Monthly sales systems told management what had already happened; they provided no indication of competitors' price changes or short-term trends and shifts in customer purchases. For IS to build a system that provided the latter capability would have involved collecting relevant information that could not be routinely derived from the sales cycle and accounting system, developing analytic and statistical techniques to assess trends and interpret patterns, and selecting suitable modeling and reporting tools for ad hoc analysis. This was often perceived as unacceptable risk, relegating it to the top left-hand cell of Figure 2-1: high business impact/low ability to deliver.

The top right-hand cell remained relatively vacant until the coalescence, in the mid-1980s, of telecommunications, personal computers, on-line industry information providers, and end-user software such as spreadsheets and statistical packages. New technology means enabled new business ends. Today, IT systems enable retailers to identify sales trends in a few days and place orders on line, sometimes several times a day, to adjust to shifting patterns of consumer purchasing. They allow airlines to analyze customer behavior and competitor pricing in real-time. American Airlines has about 120 analysts monitoring competitors' price changes, passenger traffic trends, and bookings on individual flights, using information that is literally up to the second.

An irony of the overhead era is that while the choice of applications emphasized low business impact and, hence, low business risk, the choice of technology was frequently driving investments of very high risk. As new technical products became available, the tendency was to look for places to use them. IS professionals were naturally disposed to greet each new technical development with, "Wow! What a great solution. Let's find a problem!" As long as IT was not part of the business management agenda, there was no basis for working in the other direction: from opportunity or need back to technology.

Putting the "T" Back in "IT and Competitive Advantage"

The competitive advantage frameworks introduced in the early 1980s primarily aimed at building management awareness of the business opportunities IT could provide and at persuading IS managers to rethink their own role and move out of an operations mindset toward one of business service. The frameworks were often largely remedial. For example, in 1979 John Rockart popularized the simple concept of "Critical Success Factors" (CSFs)—typically, 6–10 factors viewed by executives as critical to meeting strategic goals.

Rockart found that few companies' information systems supported their CSFs. Most "management information systems" processed largely internal accounting-based and historical information. A firm whose CSFs included aspects of competitive monitoring, human resource priorities, or research and development too often had no MIS that addressed any of these, but instead had reams of detailed accounting data. The CSF approach contributed a different perspective on IT priorities.[1] Other frameworks played similar educational roles. They emphasized choosing applications of IT on the basis of their potential business impact, moving IS up from the bottom row shown in Figure 2-1 to the top one.[2]

But though these frameworks alerted management to a relatively new and easily overlooked ingredient in competitive positioning, many of the more widely circulated models neglected the issue of matching the choice of technology to the targeted opportunity. In their zeal to convince IS managers that they needed to start thinking in business terms, many of the framework builders entirely ignored the technology itself.

American Hospital Supply's march to dominance is one of the central stories of the IT movement, yet an informal sampling of leading academics and consultants carried out as part of the research for this book revealed that only about 5 percent knew what technology base the company had used. Very few books and articles on IT and competitive advantage show any familiarity with the technology base of the most widely cited and widely taught exemplars.

The implication of this lack of discussion of technology by writers and the ignorance among the teachers is that the choice of technology is not important. "Choose whatever you fancy," they seem to be saying. "Any of the major vendors' tools can do the job and it doesn't really matter which you pick, except to the extent you care about price, ease of installation, vendor support, or other firms' experiences."

It is time to put the "technology" back in "information technology and competitive advantage." The technology matters immensely, and strongly affects the nature and sustainability of competitive advantage.

The Search for Competitive Advantage

A continuing theme in the "IT and competitive advantage" literature has been how single firms have exploited widely available technology to take competitors unawares, gain fairly immediate customer acceptance, obtain a window of competitive opportunity, and force an entire industry to play catch-up. The basic sequence of events has been: stimulus for action; first major move; customer acceptance; catch-up moves; first-mover expansion moves; and commoditization. Examples are shown in Figure 2-2.

The *stimulus for action* may be internal, for example, the search for product and market differentiation, a senior-management business vision that opens up new avenues to apply IT, cost pressures, and so forth. Alternatively or additionally, stimulus may be provided by outside forces such as deregulation, technology, industry trends, and so forth.

A *first major move* occurs when one firm, possibly following tentative and pilot moves by other firms, launches a sustained initiative. A first major mover incurs a serious and very visible investment and commitment—and risk.

When *customer acceptance* favorably affects a first mover's revenues, margins, market share, or other key elements of the customer relationship, it forces competitors to take action. But if customers fail to respond, another great idea sinks and the first mover is not added to the list of leaders that used IT to gain a competitive edge.

Catch-up moves made by competitors may include developing a comparable capability, acquiring a firm that has one or that can provide the technical base for developing it, establishing joint ven-

FIGURE 2-2 **Successes and Failures in the Electronic Marketplace**

Initiative	Automated teller machines	Home banking
Stimulus	Cost structures of branches; pressure on margins	Successes of ATMs and corporate cash management systems; perceived large market of personal computer users
First Major Mover(s)	Citibank (1976)	Chemical Bank (Pronto system) (1980)
Customer Acceptance	Rapid and consistent; convenience the draw	Minimal; no player in U.S. or Europe ever established a critical mass of customers. Many entrants to the market dropped out, as did Chemical, in 1989.
Catch-up Moves	"Shared access" networks (Cirrus, Monec); bank-specific networks, regional bank joint ventures	Mainly small-scale pilots and market tests. 19 American banks entered and abandoned the market, 1984–1989
First-Mover Expansion	Expanded locations in New York and other states; kept other banks from adding their ATM cards to the Citibank electronic franchise	Pronto abandoned in 1989
Commoditization	Strategic necessity by 1982	Already complete, even before the market is established. No unique delivery base. 15 banks in 1990 were offering services through the Prodigy personal computer-based system. Still no evidence of a real market.
Comments	Highlights dilemma of cooperate vs. compete. Consumer pressures for shared access plus operating costs of own networks forced expensive retrofit of systems. Cooperation earlier would have been cheaper for many.	The classic instance of the unmet potential: no technology blockages, but no self-justifying benefits seen by target customers

tures or forming cooperative groups to share resources (such as an industry value-added network), and using price cutting or incentives to win back customers.

First-mover expansion moves exploit the initial advantage of "occupancy" to add features, locations, or customer benefits that fortify the first mover's lead and compound the difficulties of others catching up.

Commoditization occurs over time, either because the capability is offered by all leading firms or by third partners or because the barrier to imitation decreases as the technology becomes cheaper to acquire.

Degree of customer acceptance and the resulting impact on revenues, margins, and profits create the initial competitive advantage and stimulate other firms to move. The competitive advantage literature does not draw attention to the many first moves that generated no competitor response, either because customers ignored them or the additional business and reputation were bought at a disproportionate cost.

The time window between first move and commoditization determines the competitive advantage to the first mover. The size of this window depends largely on the nature and cost of the technology base and on the management process for IT. In rare instances, the first mover may have some proprietary advantage, such as owning information that enables it to establish a monopoly. Antitrust law, rather than technology, is the principal recourse for competitors here.

No advantage can last forever, but three to seven years can constitute a valuable business springboard, especially if the first mover can add new features and incentives to the original base while the laggards struggle to catch up.[3]

The three determinants of advantage are:

1) How quickly and easily can the business idea be copied or countered?
2) How quickly and easily can the technology base be duplicated or acquired?
3) What is the likelihood that customer pressures, costs, or industry practices will quickly force companies, including the first mover, to cooperate and share technology bases?

A fair amount of evidence suggests that very few of the business ideas that have provided the basis for the competitive advantage could not be quickly copied, especially since most of them relate to basic aspects of service and core business drivers. The technology base simply was not duplicated fast enough to prevent the first mover from building a customer base that was hard to displace.

Customer acceptance clearly determines both the impact of a first mover's initiative and the need to catch up. The required technology base largely determines catch-up time and how quickly the service becomes commoditized. In both instances, time defines advantage and disadvantage.

In several first major moves, the stimulus was technology and a firm's apparent belief that it could create a new market by creating new demand rather than by meeting existing demands in a new way. Chemical Bank's Pronto home banking system is an expensive example that did not gain customer acceptance despite years of effort. Nor have customers responded enthusiastically to videotext services such as Trintex and Prodigy, despite the logical attractiveness of the products.[4]

In contrast, automated teller machines (ATMs) gained rapid customer acceptance, because the demand they created was not for ATMs and technology but for convenience. Similarly, Reuters' huge success with its Monitor service reflected not just customer acceptance but urgent customer need, resulting mainly from the volatility of foreign exchange markets in the 1970s. ATMs have become commoditized, largely through the development of systems and networks shared by many banks. Reuters has been able to resist commoditization and countermoves by its few competitors, such as Telerate, because of the cost of imitating both its worldwide IT delivery base and its ability to gather and manage data in real-time.

In corporate cash management systems, commoditization was speeded up by the availability of personal computers. Early systems (Citibank's CitiCash Manager and Chemical Bank's Chemlink were first major moves) were built on "time-shared" technology, which linked computer terminals to large mainframe computers. This entailed complex development and high telecommunications costs. Today those systems can be replicated cheaply using personal computers, spreadsheet software, and lower-cost, more reliable communication links. The entry fee for home banking has been reduced in the same way, though there is still no evidence of widespread consumer demand.

Companies and individuals were unaccustomed to buying IT-based services until the late 1980s. Knowledge of how to assess such services, judge price and quality, or handle the often difficult process of learning and adapting to them was limited. An inexperienced customer base coupled with inexperienced providers made it hard for firms to judge likely market response or estimate price elasticity. Many of the market failures during these years were not dumb ideas, just ideas that customers did not accept.

Selecting opportunities to use IT as a major element in competitive positioning is still an art form. Indeed, one question managers exposed to the success stories often ask is, "How much of this was luck?"

The Mother of Competitive Advantage: Necessity, Not Opportunity?

The widely publicized stories of companies that exploited IT effectively were striking in their overall implications. They became the theme of many articles and cases used in management education programs, which sent a clear message to IS and business managers: "Wake up! Information technology is changing the rules of business—look at this example." But they were too few in number and lacked sufficient depth and consistency of measures to provide a basis for generating reliable general principles. They were also easily overhyped and, later, easily challenged.

A flood of evangelical books and articles has preached the wonders of IT in recent years, for example, "proving" that IT provides returns on investment of 59,900 percent and claiming that it will be the central source of competitive edge in the future. This is info-hype. Claims for IT's contribution to new levels of profitability and competitive edge are largely anecdotal. For example, a number of commentators and teachers had claimed for Mrs. Fields Cookies, one of 1987's favorite citations on the conference circuit, a direct linkage between distinctive uses of IT and its economic success. Yet, just a few years later, the company was in a perilous financial position for reasons entirely unrelated to its IT investments.

Similar developments in other firms cited as evidence of IT's potential to provide massive sustained advantage have made it clear that information technology is just one element, albeit often a strong one, in competitive positioning. It cannot compensate for fundamental weaknesses in market strategy, cost structure, and organization. This is illustrated by the case of Merrill Lynch's Cash Management Account, a dramatic IT innovation that enabled the company to build a new customer base of a million affluent individuals but did not keep it from drifting badly in the following decade because of failure to manage its cost base.

If the early proponents of IT as a competitive weapon erred on the side of overenthusiasm and reliance on too small a sample base, contemporary "infopessimists" are engendering undue skepticism about the competitive value of IT that many business executives are coming to share. These doubters make many telling points, but much of their analysis rests on the assumption that as IT increasingly becomes a commodity, all major competitors in an industry can quickly converge on the same systems and services, eroding any initial advantage that might have accrued to early deployers.

This line of reasoning suggests that competitive necessity, not opportunity, should drive IT investments, and that the main impact of IT innovation will be to drive up the cost base of all players and, probably, improve the level of customer service, albeit without directly benefiting the service providers in the long term.

Many of the recommendations in this book are based on three important counterpoints to this line of thinking.

1) The management process for IT is not a commodity; management decisions, policies, and dialogues provide much of the value-added component for IT investment.
2) Technology may be a commodity, but the corporate IT platform is not.
3) Calling attention to the leaders that have not gained competitive advantage should be balanced by acknowledging the losers that are no longer in business.

The counterview that IT is a commodity and there is no real advantage to be gained from it is illustrated in studies done by several professors who analyzed in detail the history of McKesson's Economost order-entry system. In the wake of the implementation of this widely cited and admired system, McKesson reduced its sales force by 50 percent and its telephone order-entry clerks from 700 to 15 and saw sales grow by 600 percent. Yet the company's market share remained the same for the decade following its industry-leading innovation, around 20 percent. Larger competitors developed comparable systems while smaller competitors were driven to the wall; the total number of firms in the industry dropped by half between 1975 and 1985.[5]

In arguing that McKesson's growth came not from Economost

but from acquisitions, the researchers overlook the fact that many of these acquisitions were made from the position of strength afforded by Economost. The industry consolidated largely through the economies of scale and distribution provided by IT. Competitors who could not afford to build the IT capability could not adapt. This highlights the competitive *disadvantage* side of IT: mistiming IT investments and misreading industry trends.

One of the writers later concluded that it is strategic necessity that drives IT innovation in business. "Getting [strategic information systems] right merely yields competitive parity; getting them wrong may force you out of the game."[6] To support his view that IT is a source of competitive cost but not of competitive advantage he cites an economist's analysis of the factors that determine who benefits from technological innovation: the innovator, one or more imitators, or customers or suppliers. He suggests that the innovator gains advantage only when it has a patent or trade secret or the innovation cannot easily be reverse engineered. "Unfortunately [for management information systems]," he observes, "patents are rarely available, trade secrets are of limited effectiveness." If he is right, the opportunities to use IT other than defensively will be extremely rare. He comments that:

> Since the key resources—data processing equipment, telecommunications capacity or automated teller machines, for example—are commodities, [the economist] would predict that under these circumstances, all competitors with similar strategies would develop similar systems and benefits such as reduced costs or improved service would be passed on and retained by customers.[7]

This is a little like saying that anyone can create a new McDonald's; the raw ingredients are clearly commodities. Yet though many fast food chains have used these commodities, often to compete directly with McDonald's, none has yet succeeded in matching it. The commodities have not been enough.

The management process is not a commodity, for fast foods or for IT. Management is the value-added element. So, too, is implementation. To call key IT resources commodities implies that purchasing them is equivalent to using them and that there exist no major blockages to implementation. This encourages the view that it is easy to adopt new technology or create major new applications. The history of systems development and integration projects clearly shows that it is not at all easy. Indeed, there is a growing sense of crisis in the IS profession about how to dramatically improve the quality, cost, and speed of systems development. In many ways, the IT-and-competitive-advantage movement ignored the columns in Figure 2-1, which distinguish between low and high ability to deliver, in order to focus on the rows, which relate to business impact. The ideal for any firm would be to find a wealth of high-impact/easy-to-deliver applications. They would have to build the delivery capability as well as find the business opportunity. That demands a combination of people and platform.

The Management Process as the Value-Added Component of IT

Both infohype and infopessimism emphasize business choices over technology choices as a source of competitive advantage. The contribution of the infoenthusiasts—those commentators with a strong belief in the opportunities of IT—has in the end been a simple one. They have shown that IT can destabilize the competitive equilibrium of an industry. Perhaps they overargued the case for the leaders gaining a sustainable advantage, but they certainly established the potential cost of being locked out. The infopessimists' contribution has been mainly to highlight the issues of industry convergence and to reinforce the issue of competitive necessity and catch-up. Both enthusiasts and pessimists arrive at the same overall conclusion, each from a different starting point: timing, not technology, should drive assessments of opportunities and necessities. This is a management issue.

We have learned over the past decade that it is not the technology that creates a competitive edge, but the management process that exploits technology; that there are no instant solutions, only difficult, lengthy, expensive implementations that involve organizational, technical, and market-related risk; and that competitive advantage comes from doing something others cannot match. If technology magically created competitive advantage for everyone, then there would effectively be no competitive edge for anyone. If innovation were easy, everyone would be an innovator. It is not easy, as evidenced by the many barriers to transforming IT from a problem to an opportunity. Among these barriers are the troubled history of IT in large organizations, particularly the limitations of the business management process; the culture gap between business and IS people; the rapid pace of technological change; and the immense and persistent difficulties associated with trying to integrate the many incompatible components of IT into a corporate platform.

Indeed, it is the very difficulty of realizing benefits from IT that establishes a competitive opportunity. Two Canadian academics strongly counter the pessimists' argument that the commodity nature and industrywide availability of IT tools eliminate any potential for advantage. They show that it is the management process that determines the impact of technology investments.

They surveyed 200 managers and staff in 25 companies conducted as part of a 1988 study of CAD/CAM (computer-aided design/computer-aided manufacturing) implementation. They found many barriers to success in implementing the same technological systems in different firms.[8] These were largely structural, human, and technical. Structural barriers, barriers to how the organization has traditionally done things, include inappropriate business criteria for investing in IT (particularly overemphasis on cutting direct labor costs), failure to perceive true benefits, particularly intangibles, and unwillingness to take political and personal risk.

Human barriers include social inertia and resistance to change, largely resulting from fear of losing power and status. The authors found technology barriers, centered on incompatibility issues arising from the historical lack of an architecture, to be widespread.

> Incompatibility created a kind of Tower of Babel effect: different computer systems used incompatible software mounted on different hardware. Communication was possible only in the most fortunate of circumstances. Much effort was wasted re-entering data that could not be transmitted automatically. Even different generations of the same system were sometimes unable to communicate with each other.[9]

However commodity-like IT might be, the architecture that removes the barriers of incompatibility and makes it possible to build a corporate platform is clearly not a commodity. If it were, most firms would have a corporate platform. Once again, it is the management process that facilitates or blocks the development of policies for ending incompatibility.

In scaling structural, human, and technology barriers, management clearly makes the difference. To use IT competitively demands management discipline, time, commitment, and skill.

The starting point for pursuing competitive positioning through IT is for business and IS managers to:

1) understand where and how IT provides the capability to change the dynamics of competition and where it is unlikely to do so;
2) spot potential winners and avoid great ideas that turn out to be dumb, wishful thinking;
3) understand what makes an IT-based competitive advantage sustainable; and
4) carefully time the decision either to use IT as part of a major business initiative or to provide the technical infrastructure for later initiatives.

Recognizing Core Business Drivers

Most of the historical successes in using IT to change the basic competitive demands of an industry, whether or not they provided the leaders with a sustainable edge, have centered around core business drivers. These are the fundamental day-to-day elements of transactions, customer interactions, ordering, production, and delivery that define quality and service. They determine the basic cost and coordination structures of the firm, and, when there is no other major differentiator, they are uppermost in the minds of customers in choosing suppliers.

One implication of this is that IT is likely to have the greatest impact on competitive dynamics in markets in which firms are under financial pressure to find new sources of revenue, reduce costs, or increase unit margins. Companies in mature or commoditized markets have to focus on basics. Imaginative firms target IT to those basics. If they succeed in finding ways to differentiate services or improve their cost structure, competitors are forced to follow them or risk significant erosion of markets and profits.

Examples of industries aggressively applying IT to core business drivers include: retailing, which is using point-of-sale and on-line purchasing systems to manage stock levels and pricing; distribution, which is employing electronic data interchange and trade management systems that handle letters of credit, trade documentation, payments, shipping, and so forth to reduce delays and paperwork; banking, which is using ATMs and cash management systems both to facilitate fast, cheap, reliable payment and to reduce the costs of brick and mortar branches; insurance, which is applying sales force automation and image technology to the management of paperwork and policy issues; airlines, which have developed reservation systems and on-line yield management and pricing systems that manage seat inventory to maximize yield and facilitate access by travel agents; manufacturing, which is using purchasing systems, computer-integrated manufacturing, and electronic data interchange to track unit costs and quality and manage just-in-time inventories; and magazine and newspaper publishing, which is utilizing satellite distribution to ensure timeliness.

IT has transformed the core business drivers in a number of these industries over the past decade, with dramatic consequences for the rules of competition. In the U.S. airline industry, for example, deregulation in combination with technology drove the development of computerized reservation systems (CRSs), which allowed a few airlines to capture passengers by capturing travel agents. American Airlines and United Airlines together garnered close to 80 percent of this market. CRSs also later facilitated real-time inventory management, real-time pricing, and real-time profit management ("yield management"). It is now impossible to understand the mergers, alliances, acquisitions, pricing, and product strategies of any airline without examining where it is positioned—and with which partners—with respect to its CRS capabilities.

The airline experience suggests that deregulation almost guarantees a change in the basics of distribution. In more and more industries, control of the product and market is strongly tied to control of the channels of distribution. In these industries, the firm that can provide electronic convenience and ease of access to services has an exploitable edge.

In manufacturing, just-in-time operations, quality improvement, and cost reduction strategies have become survival factors for many U.S. firms. Information technology cannot guarantee, but does enable, progress on these fronts. Westinghouse, Hewlett-Packard, Motorola, and Xerox are among the firms that cite computer-integrated manufacturing (CIM), electronic data interchange, and computer-aided design systems as key elements in retooling their organizations. CIM positions firms for time-based competition, much as CRS allows airlines to pursue real-time inventory seat management.

To date, applications targeted at the basics of business have had greater impact on manufacturing than have applications in the glamour areas of manufacturing IT such as robotics and expert systems. Rethinking core operations rather than automating the status quo and using IT to achieve differentiation through service, speed, and convenience are where IT has proved most relevant for manufacturing. The same pattern is apparent in financial services and airlines, two of the earliest industries where leading firms tried to use IT explicitly as a major force in competitive repositioning. Few of the most exciting "innovative" applications worked, but those that targeted the basics of business redefined the industry over time.

Banking in the 1970s was a relatively low-technology environment. Today, to not have ATMs is equivalent to not issuing checkbooks. Are ATMs an innovation or a renovation and redefinition of service? They quickly became a competitive necessity, and are now a central element in banks' efforts to restructure costs and service. Corporate cash management systems, automated letters of credit, and electronic funds transfer systems similarly use IT to renovate well-established and basic payment systems.

A striking feature of even the most successful uses of IT that provide at least a plausible claim of competitive advantage is how rarely they merit the easily abused term "strategic," which, employed after the fact, usually means "this turned out to be very important," and before the fact, "this is my idea" or "I want a big budget." None of the innovations that reshaped the basics of various industries in the 1980s was strategic in the sense it fostered revolutionary new products or services. Each was an evolution of core transactions. Evolution became revolution when customers voted to shift their business to an industry leader that had captured the high ground, and the industry was changed forever.

Spotting Potential Winners: The "Braudel Guideline"

Why was the ATM a success but home banking a flop to date? Both address central elements of the customer-bank relationship. Both aim at improving service. Why did portable fax take off so quickly but expectations for electronic mail via personal computers remain largely unmet after a decade of sustained effort? Electronic mail is convenient and cheap. Anyone who has a personal computer can subscribe to a service such as MCI Mail for a few dollars a month. It provides many useful features, such as access to broadcast services, electronic bulletin boards, and the Dow Jones news service. Yet electronic mail is used by well under 1 percent of personal computer users. Why, too, has customer demand for so many consumer-focused information services, such as videotext, similarly lagged behind forecasts, year after year and venture after venture?

The sum of these questions is, "How can we identify opportunities that we are reasonably sure customers will respond to with enthusiasm?" One clue to the answer is the consistent pattern of IT successes that relate to core business drivers. These same drivers, *from the perspective of the customer,* whether a consumer or another organization, relate to what French historian Ferdnand Braudel summarizes as the essence of historical progress: "Development is driven by changes in the limits of the possible in the structures of everyday life."[10]

Just about every application of IT that has been widely adopted has changed the limits of the possible in the routine, essential activities of everyday life. This is true for consumer and corporate applications. Innovations that fail to gain a critical mass of acceptance and use generally do not relate to the structures of everyday life.

Consider home banking in the light of the ATM. The latter has changed the way individuals manage their daily activities. ATMs dispense money on Sunday, are on the way to work, are at the supermarket, and so forth. Home banking looks much like the ATM in terms of business concept, and in many ways is an extension of the ATM. It puts a banking facility in the house instead of on the street. Yet it barely affects the limits of the possible in everyday life. It does not dispense cash. The services it provides—checking payments, shifting funds between accounts, balancing checkbooks, and managing finances more efficiently—are extensions from everyday life.

Many of the services logically *ought* to appeal to customers. They have not. Customers see no single self-justifying (i.e., requires no elaborate explanation of why it is useful and valuable), self-explanatory (i.e., one does not have to be walked through what it is) benefit. It is not home banking technology that has so far failed to capture consumers' interest; it is that providers of home banking have not found the one simple, routine, self-justifying, self-explanatory use that will make it take off. Adding more flash to the technology or increasing the number of "useful" services is unlikely to do the trick. If home banking does take off in the coming decade, consumers will decide where and why.

A similar pattern has been played out in travel-agency versus home airline reservation systems. Conceptually, home or office CRSs should be very attractive to business people, yet growth has been modest. Most managers continue to rely on the corporate travel office and travel agent. Self-reservation systems, like home banking, do little to change the limits of the possible. The physical ticket must still be delivered or picked up. Furthermore, the volume of information and complexity of processing involved in finding connecting flights, pricing multisegment trips, and locating the lowest available fare make the systems difficult to use. What if a new home reservation system actually produced, or eliminated the need for, a paper ticket (possible with "smart card" technology) or offered a 10 percent discount? If one of these was self-justifying and self-explanatory, the calculus of benefit versus effort might change, and change quickly.

The message is clear. Technology must enable something that people want and need, not that they should *logically* want and need. The evidence suggests that an IT-based service that pushes on the boundaries constraining everyday activities is likely to be taken up quickly. If only one source provides it, that company can expect to gain a distinct competitive edge, an edge that can be sustained for years if the technology base is such that competitors cannot catch up quickly.

However "innovative" a new service, and however powerful and logically attractive its features, it is unlikely to take off unless some change in the business or social environment affects the structures of everyday life and the limits of the possible within them. Often we discuss this in terms of convenience, familiarity, ease of use, and comfort. Telex, for example, though primitive by comparison, has not yet been displaced by electronic mail because it still meets many users' practical concerns in everyday life, in a way that is "good enough." The features offered by IT-based alternatives must be more than "special" to get people to go through the adjustment, new learning, and cost of changing when existing systems are adequate.

Understanding What Makes IT-Based Competitive Advantage Sustainable

Perhaps the most difficult question to answer in considering large-scale investment in IT is, "What makes an advantage created through IT sustainable?" Several of the early and best-known competitive uses of IT in the 1980s were believed to have given the firms that created them many potential years of advantage. Yet the firms later faltered.

By definition, sustained advantage or disadvantage can only be created by means of a barrier to displacement or imitation, or through a significantly superior cost and quality base. Being first with a system that is good enough can create a barrier to displacement if customers adopt it quickly and integrate it into their everyday lives. Experience with distribution and airline reservation systems suggests that the advantage of "occupancy" can be real and sustainable for up to seven years.

With IT-free competitive vacuums becoming increasingly rare, making occupancy less feasible, firms today are more likely to concentrate on augmenting or improving operations and activities that already include IT use. But if the success is built on an integrated IT platform, more than a few competitors will not be positioned to imitate the application. Examples include: the manufacturer that can combine links to suppliers and customers for orders and delivery plus financial payments and credit services and inventory information (unless they have a comparable platform, others may be able to provide one or the other but not together); the international bank that can add a new trade financing service to its telecommunications delivery base at low incremental cost versus the competitor that has to add the cost of creating, or renting, the delivery system; the insurance company that can access customer information across products, providing a single point of service and offering an integrated service relationship; the airline that can link its reservation system to any major point-of-sale capability or hotel or car rental reservation, credit card authorization, funds transfer, or cargo or freight forwarder system; the retailer whose purchasing, inventory, point-of-sale, and credit card systems can interlink and share data to provide a real-time picture of the business.

In general, the advantages that firms create and hold through their IT investments are derived from three sources: (1) getting there first with a good enough product and skimming the market; (2) having the necessary technology platform in place; and (3) having competitors whose management does not see any compelling business reason for the platform.

Timing the Decision to Use IT

Timing the use of IT involves identifying where and how competitive, economic, and social trends might make IT a major enabling force over a two- to seven-year period. There are several reasons for this time frame.

Two years is the minimum lead time for all but very small-scale IT applications (which are unlikely to have a major competitive impact). Buying hardware and packaged software may take only a few months, but planning, developing, testing, and installing applications, as well as training staff in their use, can take far longer than impatient business managers might suspect. For major systems developments, two to five years is typical. Thus, a one-year horizon is already too late, and is a major explanation for how reactive planning leads to competitive disadvantage.

The lead time for major business innovations that depend on a comprehensive IT platform is close to seven years, an estimate supported by a number of widely cited, documented examples of using IT to gain a clear competitive edge.[11]

The rapid pace of business and technological change makes it foolhardy to plan beyond the seven-year horizon, though it makes sense to monitor long-term trends in both business and technology.

A general formula for IT timing is:

> Lead time for   = Time to build + Time to build
> launching an   the platform   the application
> initiative

The more an innovation depends on major investments in an integrated technical infrastructure, the longer the lead times. If a firm can use its existing infrastructure as the enabling element for business innovation, it has already found a source of sustained advantage—the relative disadvantage forced onto the competitor who does not have such an infrastructure and who thus faces a long lead time before it can respond either effectively or efficiently.

A firm that moves to gain a competitive edge through IT has chosen to lead rather than follow the pack. This move exposes it to a range of risks.

Market concept risk refers to the sole criterion by which an idea that seems sound and logically ought to succeed in the marketplace will be judged—customer acceptance and continued use.

Technology risk is incurred when a technology is new or unproven in the context of intended use, users, volumes, and performance requirements.

Implementation risk refers to problems in software design, project management, component integration, or vendor delivery that might impede turning an idea into a reliable product or service using an otherwise sound technology.

Economic risk is the potential for the outcomes of an effective business concept, technology, and implementation to be other than what was optimistically forecast in terms of revenue, direct or indirect cost, or support.

Organizational risk refers to the possibility that a technically functional innovation will fail to secure buy-in, threaten key aspects of the firm's traditions, norms, management processes, or culture, or require skills that the organization does not possess.

Regulatory risk refers to the potential for what appears to be a reasonable business use of IT to be blocked by regulation, governmental policy, social debate, or interest groups.

Firms that choose to lead increase business concept and technology risk and probably implementation and economic risk, since followers can learn from the leaders' experience. Firms that do not choose to lead, or that choose to follow, risk being pushed into competitive disadvantage. "When to lead—when to follow?" is the core question for competitive positioning through IT. However promising a business idea or exciting a technological opportunity, the lead versus follow question must be answered up front.

The lead versus follow question opens up a series of subquestions to be asked and answered in a specific business context. To make them more meaningful, it may be helpful for managers to choose one of the items listed in the introduction to this book as the business realities of the 1990s and answer each of the questions for their own firm. The items were: 25–80 percent of the company's entire cash flow being on line, electronic data interchange as the norm in operations, point-of-sale and electronic payments as an element in every electronic transaction processing system, image technology as an operational necessity, companies being directly linked to major suppliers and customers in electronic partnerships, reorganization frequent and not exceptional, and work increasingly location-independent.

The management questions for competitive positioning are:

1) How might, or does, the relevant technology or trend represent a competitive opportunity? For whom?
2) Does it change the core business drivers of our firm? Our industry?
3) What are the main risks associated with this opportunity: market concept, technical, implementation, economic, regulatory, or organizational? What is the nature of each of those risks?
4) Are there any indicators that this combination of business end and technology means will become a competitive necessity within three years? Five years?
5) Is the necessary corporate platform in place? Do any competitors have an equivalent platform? Will the innovation require

major changes to that platform? What is the required lead time?

6) Can this opportunity be pursued alone, or is it advisable to look for partners? Are industry partners or partners from other industries (e.g., suppliers, vendors) appropriate?

These questions seldom have clear-cut answers. If they did, there would be little competitive opportunity for anyone and firms would move in lockstep. That business managers ask these questions and struggle to answer them is crucial to the quality of the decisions they make regarding whether to strike out ahead of the pack or run with it.

Principles for Competitive Positioning

Practical management principles are gradually emerging out of the experience with and progress of competitive IT applications over the past ten years. They begin with considerations close to the firm, and extend to encompass customers and other firms within and across industries. Briefly, the principles are as follows:

1) Target IT investments to core business drivers, which are the often humdrum aspects of customer service, coordination, and costs that are at the core of opportunities for product and market differentiation. Small differentiations in core business drivers, especially convenience, speed, and ease of access to service, can change an entire industry.

2) Examine those core drivers from the customer's perspective. This should be common sense, but because the roots of IT application are in internal clerical processes, a tendency has developed to focus IT planning either on the technology or on the firm's operational needs. Too often this results in products and system features that customers "ought" to want, but, in fact, don't.

3) Assess options in terms of the time it is likely to take a competitive opportunity to become a competitive necessity, taking into account the ease or difficulty of catching up. The decision to lead or follow competitors will rest on that assessment. Lead time to duplicate an initiative and catch up determines the degree of sustainable advantage. Conversely, if the rest of an industry quickly converges on a comparable set of services through IT, the leaders may have paid a disproportionate cost to gain a short-lived edge.

4) Look for sustainable advantage to come from the IT platform

rather than from specific IT applications. Complex technology infra-structures that combine extensive reach and range of services are far harder to duplicate than individual IT-based initiatives, particularly ones that use off-the-shelf software.

Finally, it must be remembered that technology choices have business consequences, and business choices have implications for technology. Technology can never substitute for lack of business vision and realism, but technical designs can enable or block later business choices.

The more managers understand how and where IT is and is not likely to influence competitive forces, the better able they will be to slip the bonds imposed by abdication and overdelegation that keep them passively signing off on or vetoing other people's decisions. They are urged by infohype to take aggressive action or inevitably lose out to others as IT changes the rules of competition, and at the same time by infopessimists to stick to investments required as competitive necessity, there being no one who is really making money out of IT in the industry.

In the face of these countervailing forces, executives must approve decisions for which they have no reliable underlying personal framework, an increasingly uncomfortable position for capable and successful managers who handle multimillion-dollar decisions in other areas with relative ease. There is no cookbook for competitive positioning through IT, but the principles outlined here can help business executives move from awareness to vision to action.

Notes

1. See John F. Rockart, "Chief Executives Define Their Own Data Needs," *Harvard Business Review* (March–April 1979), pp. 81–93.

2. There are now so many competitive advantage frameworks that a new academic industry of frameworks for frameworks has grown up. Among the more influential frameworks have been ones built around Michael Porter's well-known and more general model of competitive strategy and of the competitive value chain. Others focus on the IT organizational learning curve as it moves through stages of development. Others offer strategic opportunity grids and checklists.

Some useful references are F. Warren McFarlan, "Information Technology Changes the Way You Compete," *Harvard Business Review* (May–June 1984), pp. 98–103; J.I. Cash, Jr., and Benn R. Konsynski, "IS Redraws Competitive Boundaries," *Harvard Business Review* (March–April 1985), pp. 134–142; C. Wiseman, *Strategy and Computers* (Homewood, IL: Dow Jones-Irwin, 1985); and Richard L. Nolan, "Managing the Crises in Data Processing," *Harvard Business Review* (March–April 1979), pp. 115–126.

3. An example of such add-ons that convert the initial product into a product stream is McKesson's Economost. It began with simple orders from pharmacists, added veterinary products and office supplies (McKesson acquired companies in these areas and added them to its electronic distribution system) and pharmacy management information systems, including patient profiles and third-party billing. It then used Economost to intrude into the traditional territory of insurance firms by providing claims processing.

4. The only major success in consumer videotext has been France Telecom's Minitel system. FT made terminals available free to telephone subscribers initially for directory assistance and then as the base for a wide range of third-party services. However, the success is debatable, given that the reported 1989 loss on Minitel was over $600 million.

5. The academic analysis of McKesson is by E.K. Clemons and M. Row, "McKesson Drug Company: A Case Study of Economost—A Strategic Information System," *Journal of Management Information Systems,* vol. 5, no. 1 (Summer 1988), pp. 36–50. There are many other detailed reviews of Economost, most of which highlight its success.

6. E.K. Clemons, "Strategic Necessities," *Computerworld,* February 22, 1988, pp. 79–80.

7. Ibid., p. 80.

8. C.A. Beatty and J.R.M. Gordon, "Barriers to the Implementation of CAD/CAM Systems," *Sloan Management Review,* vol. 29, no. 4 (Summer 1988), pp. 25–33.

9. Ibid., p. 32.

10. Ferdnand Braudel, *Capitalism and Civilization* (New York: Harper & Row, 1985). How to assess business processes from the *customer's* perspective and apply the Braudel rule in systems design and specification is addressed in Chapter 8.

11. See P.G.W. Keen, *Competing in Time* (Cambridge, MA: Ballinger, 1988).

Chapter 3

Geographic Positioning through Information Technology

It should be no news to managers that business is becoming increasingly globalized. Indeed, it is likely that there will soon be no purely domestic firms with sales of $500 million a year; any such company would surely be snatched up by a foreign company or U.S. transnational. A firm that is in an industry where the Japanese are not a threat is probably in the wrong business.

The two major catalysts of globalization are (1) the Far East's challenge to just about every aspect of U.S. business, and (2) the European Economic Community (EEC) 1992 Single Market initiative. These forces have fueled an irreversible process of cross-border operations. In a world awash in mergers, acquisitions, alliances, and joint ventures, the sudden shattering of the monolithic structures of Eastern Europe can only increase uncertainty and its shadow, opportunity.

IT has a fundamental role to play in all of this. By the end of the century, about 60 percent of all jobs in industrialized countries will owe their existence to telecommunications and computers.[1] Expenditures for the equipment and services that form the enabling infrastructures for these jobs add up to hundreds of billions of new dollars. Telecommunications is becoming every bit the major national resource that the automotive industry has been, with the result that acquisitions, imports, licensing, and contracting are political as well as business issues.

For example, Japan and a number of European nations are publicly committed to allowing foreign suppliers a fair chance to bid on major telecommunications projects and are just as clearly privately committed to keeping them out. The stakes are large. The worldwide information technology market is growing at 7 percent a year. In real terms, that means it will almost double in the coming decade, growing from close to $500 billion to almost $1 trillion. In no other major area of manufacturing and services is that likely to be true across the globe. Balance-of-trade deficits and surpluses will become a highly visible economic and political issue in many countries. The EC constitutes close to 20 percent of the global IT market, yet its IT providers have just 6 percent of that, half what they had in 1980. In 1989, four Asian cities—Tokyo, Seoul, Singapore, and Taipei—together spent $30 billion on telecommunications switching equipment, close to a third of the worldwide total.

The United States is at a distinct disadvantage relative to Japan and Europe in having no national policy for public communications networks. Public awareness that telecommunications networks are a national infrastructure as much as the railroads ever were is lacking, and not only in the United States. In Europe, telecommunications is not politically sexy—yet. When each 1 percent gain or loss in the international balance of IT trade amounts to $10 billion, telecommunications replaces car manufacturing as a national concern and potential political football. The same players will be in both games. Will the United States require Japan's IT equipment makers to agree to voluntary quotas? Will Europe reluctantly but firmly combine internal liberalization with external restrictiveness?

Over the next 10 years, more and more of the technical problems that have dominated international telecommunications will disappear. The political ones are likely to remain and even grow. The world *will* be wired.

The New Global Business Playing Field

U.S. firms have bought into Europe and European firms into the United States at a rapid rate since the late 1980s. Japanese companies, fearing a Fortress Europe, have for the first time been pushed by strategic necessity to establish major manufacturing facilities in Europe. In 1972, there was one Japanese manufacturing operation in Britain; now there are more than a hundred. The United Kingdom has been particularly attractive to Japanese business because of the

Thatcher government's policies and Britain's low manufacturing labor costs, which are 80 percent of U.S. labor costs. Japanese labor costs, which were 30 percent lower than those for U.S. workers, are now virtually the same, and German and Scandinavian labor costs are 25 percent higher than those of both the United States and Japan.[2]

The 1992 EEC Single Market initiative has stimulated aggressive cross-border acquisitions and mergers throughout the European community. This, in turn, has brought new players into the EEC, the United States being a tardy but growing force. Finland, Sweden, and Switzerland, not members of the EEC, have aggressively acquired foreign firms. The annual number of mergers in Europe has more than doubled since 1984 and investment in U.S. firms by European firms has kept pace, with France and Britain leading the spending. Until mid-1989, European merger and acquisition activities in the United States were greater than equivalent U.S. activities in Europe. The European rush to ensure a strong position in the United States is reflected in the growth of foreign buyouts to nearly $50 billion in 1988 from $18 billion in 1987 and just $10 billion in 1986. Britain spent $9 billion in the United States in 1988. In the first half of 1988, European firms spent three times as much buying into North America (350 acquisitions) as U.S. firms did in buying into Europe.

Activity *within* the EEC has been particularly intense. In the first quarter of 1989 alone, British firms acquired 80 EEC companies at a cost of $1.25 billion (mostly small ventures with an average price of $15.6 million). Britain has been followed closely by U.S., Italian, French, and Swedish firms. Much of this is small-scale jostling that will not make many headlines, but it is important to realize that the bulk of cross-border trade flow is between small- and middle-market companies and that nearly 25 percent of U.S. exports fall into this category. Future growth opportunities in trade financing, international telecommunications services, and electronic data interchange may well lie in the medium-sized business market.

Germany, which differs markedly from the United States, France, and the United Kingdom in terms of its high percentage of medium-sized firms, is an anomaly in this context. Though it has by far the strongest economy in Europe and the Deutsch mark is today one of the three leading currencies in the world, it has made relatively few cross-border deals.

Germany is where the economic power in Europe lies today. Already strong, the former West Germany seems sure to become more so, if it is able to absorb East Germany without economic strain. (One experienced German official expresses his country's general optimism concerning this issue by describing German unification as just a leveraged buyout.)

Japan, in not having been a major player in the acquisition game, is, like Germany, an anomaly. Fearing strong anti-Japanese moves by EEC regulators, Japanese firms are beginning to change their ways of handling foreign subsidiaries. A number are at last providing their European and U.S. subsidiaries' management teams with some degree of autonomy, instead of forcing them into the Tokyo straitjacket.

The Asian Tigers—Hong Kong, South Korea, Singapore, and Taiwan—are concerned that their export-intensive economies may be locked out of the EEC unless they establish a presence within its borders and have become new players in Europe. Firms and cities in these countries also see the importance of IT in managing international trade shipments. Singapore is a world leader in electronic data interchange, and Singapore and Taiwan have built a strong supply of well-educated, fairly young workers in computer programming and computer science.

In Europe, the roles of countries such as Spain have changed dramatically. Once a backwater of investment, Spain has become a major target for mergers and acquisitions. Recognizing that transforming its telephone system is an imperative for national economic development, it has enlisted major U.S. and European telecommunications providers to completely renovate what is among the worst phone service in the developed world. Hungary similarly decided to privatize its telecommunications monopoly while still under the old Communist regime, before it had any expectation of recovering its independence.

European business was already acting in 1990 as if the 1992 Single Market initiative was a fait accompli. European firms were making aggressive and often cooperative moves, for example, developing industry and cross-industry value-added networks to handle electronic data interchange, electronic shipping, insurance, and other services that make an IT base a geographic necessity for being a major player in many markets.

IT and the New Global Playing Field

Globalization of business demands globalization of organization. Two extremes of international organization are the "multinational operations base" and "transnational platform." Figure 3-1 highlights the advantages and corresponding disadvantages, particularly those of a political nature, of each strategy. (The distinction between multinational and transnational firms comes from the work of Bartlett and Ghoshal, whose book *Managing Across Borders* provides many detailed illustrations of firms along the spectrum.)[3]

The Multinational Operations Base

A multinational firm is composed of relatively independent national and regional units, upon which corporate management imposes only limited requirements for worldwide commonality in key processes and operations. In such firms, management of IT is usually local, excepting shared international telecommunications networks. Planning aims at ensuring just enough coordination to meet international communication and control needs, but avoids global planning that might be construed by national business units as an intrusion on their autonomy.

A product of a less frantic time, the multinational is ineffective in highly interdependent markets and operations and it founders under the speed of business change and international competitive pressures. In the old days of long development and rollout times, innovative U.S. firms such as 3M could launch a new product in the United States, test market reaction, and then gear up to launch the product in Europe. Foreign firms like Canon increasingly exploited this time to learn from 3M's U.S. activities and get its own product into Europe first.

Firms have found it hard to adapt the old multinational form to the new business context. Many new coordination mechanisms have been tested and found inadequate. Matrix-based management structures that combined intracountry line responsibility with cross-country functional responsibility and task forces set up to handle shared planning rarely worked. Old fiefdoms blocked coordination, and new organizational units added complexity. The push toward greater collaboration demanded major shifts in planning and control, hence in communication and information flows. Because the time windows between many countries are small, new telecommunications highways are needed to make communication easy, fast, and effective. To handle forward planning in an environment of regional or central manufacturing and local distribution plus shortened cycle times requires adequate information systems and telecommunications capability for moving information to and from the field. The traditional multinational model is obsolete in this context.

FIGURE 3-1 **Multinational versus Transnational IT Strategies**

	Management Strategies				IT Strategies		
	Structure	Priority	Historical Analogy	Central Coordination	Telecommunications	Transaction Processing	Systems Development
Multinational Operations Base	Largely autonomous units, at level of country and/or region	Local response to national market demands and conditions	British Empire—colonial model	Mainly through the capital budget and reporting systems	Mainly individual business-unit and country/region decisions; some shared "backbone" networks, especially U.S. to Europe	Very limited needs for international systems; finance the most likely exception, with worldwide cash management typical	Separate units matching country/region structure and key locations
Transnational Platform	Selected key functions coordinated worldwide with equal emphasis on local autonomy and responsiveness, such as global product development with local marketing strategies	Entire business system optimized; operated as if there are no national boundaries	Federal system—national government plus state rights	The main challenge: how to coordinate without intruding or over-controlling	Platform essential, with substantial reach across international locations and range across key business functions	Transnational processing systems an essential part of the business strategy	New cross-functional and cross-location planning mechanisms and collaborative teams

The Transnational Platform

For the transnational firm, countries and regions are like states in a federal system. Corporate management is tightly coupled in many areas of the business, including global manufacturing, product development, coordination of R&D, and marketing. Such a firm will treat IT as a resource to be coordinated worldwide, necessitating a common technical platform for telecommunications, information management, and many aspects of other IT services and operations. The network is akin to the interstate highway system, which allows variations in state driving regulations. There are worldwide reporting and processing systems, analogous to the federal income tax system, complemented by state-specific systems. The federal structure is not a monolith, but a flexible combination of federal and state systems, a balance of centralization and decentralization (see Figure 3-2).

Firms face many difficult technical, regulatory, political, and organizational challenges in justifying and implementing the transnational IT platform that makes practical the transnational organizational structure and coordination. The technical challenge is associated with the availability, differences in components, and costs of the technology across the world. Regulators often insist that firms use national equipment suppliers, impose heavy cost penalties on business uses to subsidize consumer services, and favor particular technical standards and block others.

Within the firm, the challenge is to establish policies and effective management processes to ensure that the IT platform provides international coordination without compromising the autonomy of national business units. The attendant organizational challenge is to create mechanisms that foster cross-cultural collaboration throughout the firm. Building the momentum and cooperation needed to create a domestic platform and policies for coordination can be difficult; extending the platform across international boundaries can verge on the impossible.

Fresh practical thinking can make the transnational IT platform a likely source of competitive advantage that will be strongly sustainable throughout most of the 1990s. Only a small fraction of U.S. firms are likely to learn how to think internationally about IT. In only a handful of leading companies committed to global business planning is there an IT strategy built on explicit policies that ensure the same level of transnational coordination for IT as for manufacturing, finance, or product development.[4] Unless this level of coor-

FIGURE 3-2   **The Transnational Platform versus Multiple Facilities**

One Firm—Four Networks

Manufacturing network for linking plants

Financial network

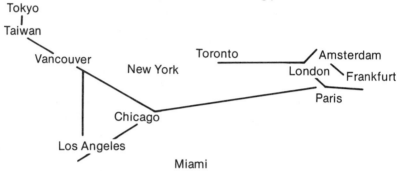

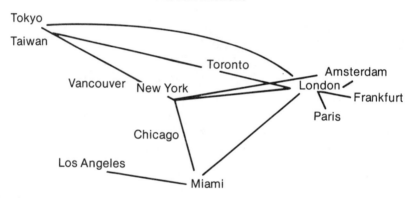

U.S. internal network for linking offices

International EDI network

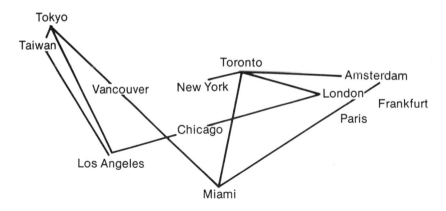

The result

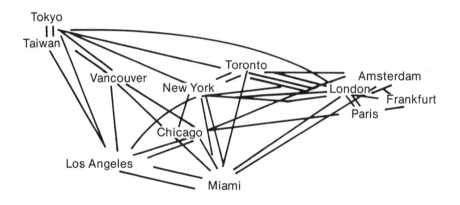

One Firm—One Platform

Multiple systems rationalized into a transnational platform

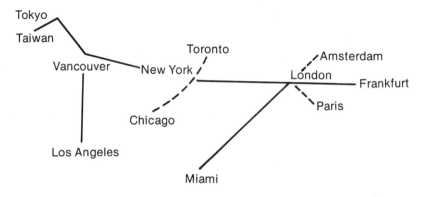

dination is a priority for senior business managers across the firm, the company will likely drift toward the multinational operations base approach.

Becoming a Transnational

Firms do not drift from a multinational operations base to a transnational platform. They have to actively make the shift. Competitive positioning through information technology is a topic well mined in books, conferences, and the business press. *Geographic* positioning, though it is where IT is likely to be a fundamental rather than peripheral issue for many companies' business future, has been largely neglected.[5]

IT is the essential mechanism for communicating across time zones. Electronic mail and fax are the only sensible ways to get messages to people from London to New York and Tokyo without either staying up very late or getting up at 3 A.M. Information, being the basic currency of coordination, must move at the same speed as business. IT speeds information. Clearly, an international business plan without an international IT plan is a contradiction in terms.

Yet few firms have such a plan. In most large multinationals, telecommunications, the key element in using IT to bridge boundaries of geography and time, is (1) managed separately in the United States, where telecommunications is fully deregulated and prices have dropped rapidly in a context of fierce competition and accelerating technical innovation; (2) handled on a regional basis in Europe, where telecommunications policy remains restrictive and services expensive, with "liberalization" of the traditional quasi-government monopolies underway but still limited; and (3) limited to ad hoc links to the Far East, where telecommunications costs, availability, and quality vary widely.

A leading manufacturing firm that announced a business strategy based on manufacturing and distribution through regional centers of excellence is an example. The firm plans to coordinate worldwide operations by giving responsibility for key activities to specified locations. France, for instance, will be the center for engineering, Canada for R&D in process control, and the United States for CAD/CAM development. Manufacturing facilities, currently handled on a country-by-country basis, will be consolidated regionally.

The firm's organizational vision is of coordination across a global market, with each country a unique contributor. Yet the company has no international IT blueprint. It operates highly efficient U.S. and several smaller European networks, and leases a number of medium-speed communication circuits to its Tokyo offices and Taiwan manufacturing plant.

Who runs the Atlantic links between the United States and Europe? Who coordinates the transnational telecommunications platform? What shared processing systems and information stores are needed to coordinate manufacturing, distribution, finance, engineering, development, and marketing? Where should the Far East telecommunications "gateway" be located? (See Figure 3-3.)

These questions have not been raised, let alone addressed, at top management levels. The head of Information Systems believes

FIGURE 3-3   **Wiring the World: Nodes and Gateways**

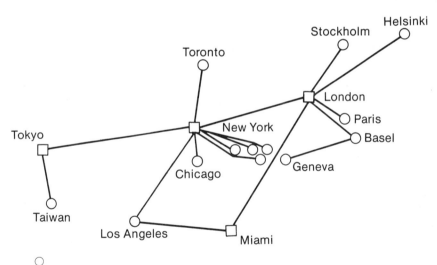

□ = Gateway

○ = Other nodes on the backbone network

The solid line indicates the backbone network; nodes connect to each other via the gateways. For example, Geneva connects to Los Angeles by connecting to the Basel "switch," which connects to the London backbone gateway. The gateway routes the message to Miami, which connects to Los Angeles.

senior management abdication will "cost the firm a fortune in the nineties."

Another example is a European firm that rationalized a substantial premium it had paid to make a major acquisition in the United States on the basis of joint efficiencies and economies of scale that stood to be realized by the combined companies. Subsequently, the senior executive dispatched to review the firm's U.S. and European operations concluded that IT costs were far too high and, consistent with the CEO's philosophy of devolution of authority to local divisions, replaced the central IS unit with divisional groups, with their own staff and budgets. He reported the cost savings and became a hero.

The savings are illusory and the hero has moved the firm backward. Already, ad hoc, application-by-application telecommunications networks and new information systems are being developed to satisfy unmet demands for coordination. Information flows are fragmented and cumbersome; corporate management in Europe does not have the information it needs to run the firm as one entity; senior managers complain that they have had to delay plans for joint product development. Stated one: "The information is a mess. It takes months to find out what has been happening. By then, it is too late to take action effectively."

Costs of data processing are increasing as special-purpose systems have had to be built at a crash pace. Every new system demands new telecommunications that has to be acquired from outside service companies. One internal study estimates that a shared United States-to-London-to-Germany "backbone" network would cut telecommunications operating costs by a factor of six. It is not a practical option for the firm, because its U.S. and foreign computer systems and information resources are incompatible with each other and have entirely different telecommunications requirements. Top management sees the problem as one of IT costs that must be brought under control. The former head of IS in Europe, who admits to failing "completely to get any one to listen to me about the need for a worldwide telecommunications architecture," was pushed into early retirement at the end of 1989.

Effective decisions about geographic positioning depend on the broader management process. In both these firms, senior management has a business vision but lacks a compelling message for an integrated IT platform. The result, as we saw earlier, is delegation of IT planning to the IS function with no mandate for making sure that there is a transnational IT plan to match a transnational business plan.

Building Transnational Collaboration

Too many U.S. firms look beyond IT as a source of competitive advantage domestically to IT as a transnational competitive opportunity or necessity. Overconfidence in U.S. leadership in IT has until recently led many of them to overlook the lessons to be learned from Europe and the Far East, the companies of which have responded to the realities of globalization far faster and earlier.

It is up to top management to ask the question: "Is a transnational platform a requirement for competitive and organizational health?" If not, a firm can reasonably minimize the extent of international coordination and commonality of standards, services, and shared resources. If it is essential, the firm must rethink the entire IT management process and act fast. Managers can be sure that a transnational IT platform is essential if the combination of the following conditions represents either operational or competitive necessities or a substantial competitive opportunity.

1) Business demands a coordinated international plan for manufacturing, marketing, distribution, and service. Piecemeal, application-by-application investment in IT when this is a necessity yields a geographic jumble and an operational nightmare.
2) Effective competition depends on time-dependent coordination across national boundaries. If it does, delays will mean missed sales, lost customers, and belated decisions.
3) Major national competitors and customers rely on electronic linkages. The global firm that tries to compete in Europe and the Far East without an EDI capability does so at a guaranteed disadvantage.
4) The time zones across which the firm's core activities are performed span more than six hours in width. Here, IT determines reaction times; a transnational IT platform is a competitive enabler, its absence a competitive disabler.

Creating the organizational, cultural, and technical base for designing a transnational platform is a challenge. Technical issues, which mainly relate to standards, are addressed in Chapter 7. Even when these issues are resolved, organizational and cultural issues are likely to remain a challenge.

In general, national units in a multinational can be expected to oppose coordination, particularly when it is imposed from the center by technical managers and staff who do not understand differences in regulation across countries and the constraints they imply. To persuade individual units to work collaboratively to build a transnational capability requires highly compelling reasons and a clear top management directive.

Cultural differences, most obviously the impact of language on understanding and misunderstanding and on cooperation and antagonism, can also block collaboration. Richard Kuehn's review of a telecommunications conference in Amsterdam in 1989 provides an entertaining cautionary tale.

> To show you how poor the communication was between end-users and some of the speakers, during one session an end-user asked a manufacturer what steps had been taken to prevent penetration by computer viruses. Because of the end-user's accent, the word "viruses" came out sounding like "wiruses." The manufacturer didn't try to make sense of the question, but instead responded haughtily, "I see no difference, nor should you care, whether information is transmitted by wires, radio, or some other means."[6]

Is the communication problem one of language or of the ignorance of a telecommunications specialist, who did not recognize the term "viruses," a commonplace one for computer professionals?

If IT is hard to plan and manage within a single country, it is exponentially more difficult to handle internationally. When senior business executives are unaware of the need to include IT in geographic positioning, U.S. telecommunications specialists are unfamiliar with foreign environments, opportunities, and constraints, and telecommunications and information systems professionals do not understand the fundamentals of one another's businesses. If there is no transnational business vision, no compelling message about the need for an integrated platform, and no IS technical vision, IT planning drifts and defaults.

International Information Systems

Most of the critical issues relevant to geographic positioning via IT relate to telecommunications, for the obvious reason that this technology is central to bridging time zones and physical distance. The telecommunications base facilitates coordination of operations, which in turn creates pressures for common information systems in specific areas. Firms that have developed a global manufacturing strategy, for example, subsequently experience a need for consistent and supportive ordering systems. These systems can rarely be identical, given different tax laws and structures in different countries, but they should provide a pool of shared information, facilitate transnational processing, and ensure smooth and error-free operations. They may be developed in one location and transferred to others, with suitable modifications, or they may be developed by transnational, and hence cross-cultural, teams. The latter will require skilled leaders.

The cosmopolitan manager who can work collaboratively across cultures is rare. To be international is not the same as to be cosmopolitan. Many U.S. managers travel to London or Frankfurt, stay in U.S. hotels, and see nothing of and learn little about the country they are in. They assume that decision-making styles in Europe are the same as in the United States and expect that because English is now the common language of business, their words, assumptions, and attitudes are the same as those of the English-speaking customers, colleagues, and bureaucrats with whom they deal.

Many firms have recognized the need for the cosmopolitan manager. The profile of the ideal "Euromanager" as defined by headhunters is someone who speaks two languages, spent part of late childhood and adolescence abroad, and has worked in several countries. This is a rare bird in business and a rarer one still in the technical IT field.

Cosmopolitanism will become more and more valued in business and IT. Telecommunications in an international firm demands it, and so too may Information Systems, and very soon. Software development has historically been a U.S. preserve, with individual national markets. Almost every major office technology and personal computer software package—e.g., Lotus 1-2-3, WordPerfect, DBase III, and Microsoft's products—were created in the United States. Software development may soon join supercomputing as an endangered American technical species.

Software has become a major international industry, with many powerful players. In Europe, strong software firms have been created by merger and acquisition activities associated with the 1992 Single Market initiative. These firms draw on the distinctive strengths of French developers in large-scale, real-time systems and disciplined British development of advanced commercial software. Its factory approach has disproved the belief that Japan lacks the creativity to produce successful software by turning out software that is cheaper and has fewer defects than equivalent U.S. software, according to a study made by a leading U.S. vendor. In the past five years, the number of Japanese software houses has increased from several hundred to several thousand.

Worldwide, countries and companies are offshoring software development to low-cost labor areas, with programmers submitting their work via satellite for compiling and testing. Offshoring software development teams may be a way for U.S. firms to compensate for a declining skill base; Budapest, Bombay, Tel Aviv, Singapore, and Dublin are targets of opportunity but only for the firm that can link them electronically to its operations centers. India, which has invested heavily in satellite telecommunications, has announced its intention to garner 2 percent of the world market for software exports. Israel, Hungary, Mexico, and other countries with a well-trained technical labor force but a weak economy and limited IT base are being wooed by U.S. firms such as Citibank, Motorola, and Texas Instruments and by well-funded Japanese investors. Finally, East Europe is likely to become a significant niche player in software. Hungary, whose lack of access to computing facilities is offset by first-rate analytic and scientific training and skills in creative programming, is particularly strong in this area.

Competing through Geographic Positioning

IT's impact on geographic positioning has fueled a number of business trends. Among them are: bringing work to people via telecommunications, instead of moving people to the work; creating regional centers of expertise and consolidating operations via telecommunications; and basing decisions about location on the quality of telecommunications services.

An important aspect of geographic positioning through IT is the ways in which cities and national governments either actively court firms that are pursuing a transnational strategy or block the efficiency and effectiveness of that strategy. Many business and IS managers are likely to be unfamiliar with how cities compete through IT. With business and IT choices and consequences increasingly interdependent in a context of global operations, this unfamiliarity can mean overlooking areas of IT planning that may significantly enable or disable business flexibility.

How Cities Compete through Information Technology

IT is becoming a salient aspect of many cities' and countries' economic policy.[7] The city of Kawasaki, Japan, is creating the "intelligent" metropolis, wired for business. In Europe, Rotterdam has built the electronic port. U.S. cities that have spotted the competitive opportunity afforded by IT include Heathrow, Florida, and Breda, Iowa. Heathrow is located north of Orlando. Heathrow city leaders' efforts to create a "city of the future" via telecommunications have resulted in relocation of the American Automobile Association headquarters from Fairfax, Virginia. When the AAA outgrew its headquarters site in Virginia, senior management had identified telecommunications as one of the main criteria for relocation. Heathrow was just one of the cities that met the other requirements—quality of labor force, housing, closeness to a major airport, and real estate costs. Heathrow's unique telecommunications infrastructure was the differentiator.

It was Heathrow's mayor, Jeno Parlucci, a real estate developer and entrepreneur, who advanced the idea of using an advanced telecommunications infrastructure costing more than $2 billion over a 10-year period to attract business. With Southern Bell's installation of a fiber optic network, Heathrow will become the first residential provider of ISDN (Integrated Services Digital Network), the long-planned, expensive worldwide conversion of the telephone system's analog voice transmission to digital voice, data, document, graphics, and computer transmission.

Heathrow's new telecommunications capabilities are the product of cooperation among three groups. Access to a 3,000-acre, greenfields site enabled real estate developers to design intelligent buildings, that is, buildings specifically configured for computers and workstations. Bell South installed a fiber optics electronic highway system for telecommunications transmission, and Northern Telecom provided specialized digital switches that manage and direct the flow of high-speed communications.

Breda, Iowa, with a population of just over 500 people in 1988, lost its high school, rail connection, and even its car dealership when farming slumped in the early 1980s. The installation of 13 miles of fiber optics in 1988 made possible the capture of the new sales center for one of the largest U.S. telemarketing firms and positioned Breda to grow, not *in* high tech, but because of it.

Nebraska's and the city of Omaha's initiatives, explicitly proposed in a 1984 report of the governor's task force ("Nebraska as a World Class Center for the Communications Industry"), constitute one of the most striking instances of using telecommunications as a factor in economic development. Omaha, because it is at the intersection of several major north-south and east-west fiber optics networks, has some of the lowest "800 number" phone rates in the country, making it attractive to telemarketing firms and other heavy users of telecommunications such as credit card firms and hotel reservation services. A sustained partnership with Northwestern Bell made Omaha the 800 number capital of the United States, with more than 100 million calls being handled by 100,000 employees of 25 companies. Fireman's Fund has consolidated its customer service telephone operations in Omaha; Northwestern Bell is moving its computer operations there; and Union Pacific picked the city for a $55 million dispatching center for expediting trains across 20 western states.

Omaha's attractiveness to telecommunications-intensive businesses derives from several factors. One is the city's emphasis on providing a skilled IT labor force. Omaha has one of the very few university programs in telecommunications management and provides job training programs for some firms' new employees, paying half the cost. The responsiveness of Northwestern Bell, which will install 800 lines in 24 hours versus 1–2 months elsewhere, is another factor. Finally, there is top officials' and policy planners' awareness, vision, and compelling message for the metropolitan platform.

Many of the firms that have relocated functions on the basis of telecommunications opportunities have consolidated operations previously spread over several locations. AAA, for example, is co-locating its Miami facilities with its new Heathrow headquarters, and Union Pacific is closing existing dispatching centers in eight states. These firms did not simply move a function from one place to another; they rationalized and brought together separate operations.

In an international context, national telecommunications policies significantly affect ease of access to and quality and cost of telecommunications. The liberalization of telecommunications encourages transnational firms to locate regional centers and back-office functions in that country. Cities that become international electronic trading and logistics centers—nodes in key electronic trading networks such as those for foreign exchange and securities—gain rapid and often overwhelming advantages over cities that, though they may be stronger economically, lack first-rate and comprehensive IT facilities.

National telecommunications policy has significantly affected how four major European cities—London, Amsterdam, Frankfurt, and Paris—compete as trading cities, particularly in the areas of financial services, cargo and freight, and attracting multinationals.

London and Amsterdam represent liberalized telecommunications regimes. The United Kingdom, which ended monopoly control of telecommunications in 1983, was the first country outside the United States to do so. The Netherlands privatized its PTT (Poste Télégraphe et Téléphonique) in January 1989. France and Germany, in contrast, have fought a rear-guard action to preserve communications as a monopoly run by a quasi-governmental agency. Germany gave up the fight in mid-1989, joining the Netherlands and the United Kingdom in accepting competition in most areas of telecommunications.

London has been a major beneficiary of liberalization, and Frankfurt a victim of the Bundespost's erstwhile restrictiveness. In the late 1970s, some international banks moved their main foreign exchange, back-office operations, and computer data centers out of Germany. Nearly half of the world's daily foreign exchange transactions are made in London, which handles about $100 billion a day. Despite the importance of Germany and its currency, the Deutsch mark, as an economic power, Frankfurt is a minor player in the foreign exchange markets, because those markets are entirely dependent on access to fully up-to-date information and speed of communication. No German bank has the equivalent of Chase's or Citibank's global network for electronic funds transfers, letters of credit, cash management, trade documentation, or similar services.

Britain is the European leader in electronic data interchange and value-added telecommunications networks mainly because Germany is the laggard.[8] The United Kingdom has used IT as a competitive force compensating for weak manufacturing and a poor gross national product per capita.

The Netherlands, like Britain, is strengthening its cities' and companies' competitive positions through IT. The Netherlands is the logistics center of Western Europe, with 20 percent of the Common Market imports and 30 percent of the exports being shipped through it. The port of Rotterdam and Schiphol (Amsterdam) airport, major clearing points for cargo, face immense challenges from

Antwerp, Le Havre, Hamburg, and the Channel tunnel (due to be completed in 1993). Rotterdam's pioneering use of IT to cut clearance times through customs to 15 minutes is part of its defensive strategy.[9]

Lufthansa, Air France, and British Airways are competing with KLM in the expanding and profitable air cargo market. Schiphol and Heathrow airports are vying to be the main gateway into Europe in the (gradually) coming deregulated international airline environment. Cargo accounts for approximately 20 percent of the revenues of Lufthansa, KLM, and Air France, and is expected to grow as manufacturers move to just-in-time production and inventory management (turning their inventories over 12–16 times a year instead of 2–4). Emery Air Freight calculates air freight will grow from $18 billion in 1987 to $29 billion in 1990 and be roughly the same size as the international business travel market by 1995.

Increasing reliance on on-time delivery has left customers increasingly indifferent to how goods travel, whether by air, sea, rail, or road. In addition, they want timely end-to-end information about shipments and less and faster paperwork. Schiphol has responded to these demands by expanding airport capacity and investing heavily in IT. Its Cargonaut system, a leader in cargo automation, is to be linked directly to Rotterdam's INTIS system, which, as mentioned earlier, allows goods to be cleared in as short a time as 15 minutes. Relevant trade, administrative, customs, financing, and insurance documentation is sent ahead of the goods in electronic form.

IT is an explicit element in the Dutch strategy to gain geographic advantage. The HERMES system handles rail freight, as does INTIS for sea freight, Cargonaut for air cargo, and SAGITTA for customs, creating a platform for building a total electronic logistics chain into which customers can link their own computer systems via EDI. By comparison, clearance takes days in most ports, and cargo spends 90 percent of the time in transit sitting on the ground.

Rotterdam, Schiphol, and Amsterdam planners all recognize the need to invest heavily in IT to maintain the Netherlands' proud role as "Distribution Center—Europe." So do many of its competitors. British Airways is spending $100 million on cargo improvements and has established automated links with its top 100 shipping agents. A group of public and private firms in the greater Paris region is forming a "Teleport" consortium to build a shared value-added network for freight that will link all major regional airports.

Communications quality and access are obviously key elements in providing services such as these. Much of the reason for the

United Kingdom being the current leader in EDI in Europe, the Netherlands a contender, and France (to date) and Germany laggards is the clear link between telecommunications monopoly and prices. In mid-1989, the prices for a basic leased circuit (9.6 kb, analog) were (per annum):

	Domestic	International
United Kingdom	$ 2,800	$12,000
Italy	6,000	43,000
France	6,100	15,000
Germany	9,400	20,000
Spain	15,500	30,000

(France, whose technology is the most advanced in Europe, has caught up in many areas of pricing since 1989, but the overall range of differences across the EEC remains as broad as shown here.)

Cost per circuit was the same for all providers: $1,600 for domestic service and $2,400 for international service. Deregulation and competition bring price closer to cost. In the United States, MCI and US Sprint have contributed to the continued reduction of the cost of long-distance phone calls. Lack of comparable deregulation to date has kept local phone calls disproportionately high in the United States. When Britain removed BT's monopoly on the supply of medium-sized telecommunications switches, prices dropped 30 percent.

No wonder banks and multinationals move their back-office operations and data centers out of Germany and France! No wonder the Netherlands completely deregulated its PTT and established the most liberal policy on the use of leased lines. No wonder Philips, the Dutch electronics giant, has set up a subsidiary to handle its expanding EDI operations. The firm has also established a consortium to develop an electronic case-handling system for the European insurance industry, which is estimated to consume 40 million trees per year for paper.

Telecommunications and Business Location

The examples of Britain and the Netherlands at one extreme, and Germany and France at the other, hint at the importance of management awareness of the impact of IT on business location. A European spokesman for Federal Express observed that "Airline products alone do not guarantee express service; the cutting edge is telecommunications," which suggests that his firm is fully informed about IT. The company's choice of Brussels as its main cargo hub reflects that awareness. Federal Express has already helped to establish Memphis as "the distribution center of the United States" (the Memphis mayor's words).

By contrast, the British bank that moved its simple international back-office management system into Paris and had to write it off at a cost of over $10 million in the mid-1980s lacked any awareness of the implications of differences in telecommunications regulation. The bank's system employed a standard transmission technique ("packet switching") over a "private" (leased line) network and used custom software. The French PTT at the time maintained total control of packet switching through its monopoly Transpac network; it was illegal to send packets via a private network. This system could not be adapted and ultimately had to be scrapped.

By and large, U.S. business executives assume that needed domestic telecommunications facilities can be obtained quickly from the growing number of competitive providers and that there are no significant constraints on applying standard telecommunications services to other components of their IT base such as transaction processing and access to information stores. Thus they leave the details to their technical staff. Transferring these assumptions to the new international context is foolish at best. The following realities are the axioms of geographic positioning through IT.

1) A transnational business strategy requires a matching transnational IT capability, both in terms of technology infrastructure and organizational mechanisms for planning and coordination. There is little natural momentum for creating this, especially in a diversified and decentralized business. The momentum must come from corporate management.

2) Even with the growing global trend toward deregulation of telecommunications, there remain many national roadblocks that, from a business viewpoint, are unreasonable. It is often illegal, for instance, to log phone calls automatically through a telecommunications switch in order to itemize internal billing for use of the firm's network; it violates privacy laws in some countries. Similarly, a country's postal/telegraph/telephone agency (PTT) may have rules that prohibit the use of specific equipment made available in the country. Given these circumstances, a global strategy for telecommunications cannot be just an aggregation of local strategies; there must be a global plan.

3) Coordinated planning and implementation across national

boundaries stretch the limits of competence and good temper. The firm's IT staff, which often has enough problems dealing with local situations, can be severely stretched by having to divert time and attention to trying to resolve transnational problems.

There remain many uncertainties in almost every area of telecommunications policy, technology, and the ability of PTTs to become market-driven service providers. Leading telecommunications suppliers, firmly committed to competition, are working hard to improve this situation. British Telecom, KDD Japan, and MCI now provide a joint international network capability. The U.S. and British governments are working to break up the cartel arrangements that make international phone calls into the United States disproportionately expensive compared with calls from the United States.[10] British Telecom acquired Tymnet, a major U.S. value-added network provider. MCI bought part of Infonet, a well-established international supplier of telecommunications services for business, whose other owners include several European PTTs. AT&T purchased a comparable British firm, Istel. Yet one-stop shopping for transnationals remains a major and unresolved problem in Europe.[11]

Similar uncertainties dominate the Far East. Japan is committed to its own form of liberalization, based on the principle of tight PTT control of basic services but total competition for value-added services. At the same time, it protects its markets from foreign telecommunications providers, while paying lip service to open bidding. Korea's new National Telecommunications Association is similarly split between deregulation and protecting Korean equipment makers from foreign competitors.

Australia is a growing area of innovation and competition. One of its banks, WestPac, is recognized as one of the top five in the world in electronic banking, and Qantas, its national airline, is among the leaders in on-line, real-time information systems. India has moved aggressively in the past few years to use satellite transmission to improve its telephone services and provide computer communications for industry and universities.

EDI is a priority for many of the Asian centers of trade. Singapore, which in just two years created the TradeNet network for electronic data interchange between Singapore's trading companies and government agencies responsible for issuing trade documents, is a striking example of a city-state that has recognized and responded to the need to dovetail economic planning with IT policy. More than

500 companies have subscribed to Tradenet, and more than 95 percent of government permits are now processed in 15 minutes, rather than what previously took one to four days. Permits are automatically routed to port and aviation authorities to speed up the physical clearance of goods. TradeNet processes excise duties, customs fees, and funds transfers.

Export-dependent Hong Kong's government, banks, and traders also recognize the need to replace trade paperwork with electronic document management. Each of the more than 100,000 trading firms in Hong Kong send between 2,000 and 10,000 documents per year. Tradelink, a consortium of 11 leading firms, asked the Hong Kong government to establish a commercial EDI firm. The general manager of one of the largest, 120 of whose 300 employees work full time on documentation, explains the business logic for this aggressive commitment to EDI.

> If you assume an average consignment is 10 tons, that means we handle about 440,000 bills of lading a year. . . . These, in turn, need to be supported by 440,000 shipping order sets—which are usually seven- or eight-part documents.
>
> Our aim at Swire Shipping is for most consignment details to reach the importer within 72 hours of departure from Hong Kong. On a regular P&O sailing to Europe, which takes 19 days, this gives the importer 16 days to plan his collection and warehousing arrangements.[12]

In both Hong Kong and Singapore, government telecommunications authorities work closely with the business community. In many others, they do not. But, regardless of the PTT-business link, telecommunications is a maelstrom of change. Every major player wants to be sure it is not left out of the action. Alliances between PTTs, U.S. telecommunications providers, computer vendors, and telecom equipment manufacturers are rumored, announced, delayed, implemented, and broken off. Many are likely to be short-term relationships. The international telecommunications world is in flux, with no promise of equilibrium or even slowing of the pace of change.

Management Action

Whether viewed in terms of how cities compete, how countries exploit electronic offshoring, or how companies consolidate operations, choose locations, or access skilled staff, IT is part of international business now. To create a badly needed new level of planning for IT, transnational firms must explicitly resolve the problem of organizational diffusion of authority, as discussed here in relation to telecommunications. Given a business vision for geographic positioning and a compelling message in support of an international IT platform, a new locus of authority becomes vital to effective coordinated planning. A guiding blueprint is needed to bind separate national systems and facilities together. The firms that have most successfully balanced corporate coordination of the technology platform and local business decision making on the use of the IT base have an international architecture and an architect, who is thoroughly familiar with international telecommunications and works in a consultative relationship with managers in every country continuously. Familiarity with only the United States is inadequate, and can even be a liability in that it encourages a naive view that the entire world must and will deregulate telecommunications and that common information and processing systems are just an issue of telecommunications connectivity.

Business management must first have awareness, then vision, and then make an explicit choice of either a platform-based or an application-by-application strategy. The strategy for management action is subsequently fairly simple.

1) Review the need for international coordination, shared processing systems, and shared information, not just for the short term but into the 1990s. In particular, examine in detail the firm's business plans for Europe and the Far East and its assumptions about the post–1992 era.

2) Establish a clear responsibility for defining the international IT platform or shared facility base(s). Ensure that there is both a firm statement of architectural principles and close ongoing consultation between a cosmopolitan central group and dispersed geographic units on both a regional and national basis.

3) Make it unambiguous which aspects of IT require central direction and which can be left to local initiative.

4) Recognize where telecommunications cost, quality, access, and reliability are essential determinants of business cost and quality, and choose locations for operational unit activities and key business activities carefully. Avoid countries that obstruct progress here.

5) Require the corporate IT group to visit the field regularly and to focus on the business as well as the technical context of its activities. At the same time, ensure that the group keeps abreast of

progress in technology, applications, and standards around the globe. Evaluate the group's awareness of EDI in Europe and Asia, of value-added networks worldwide, and of international software industry growth.

6) Fund the international telecommunications backbone network as a shared business asset.

7) Build knowledge of international IT issues among all technical and business units. Emphasize the business and organizational opportunities and necessities of geographic positioning via IT.

8) Encourage representation on international user committees concerned with standards and policies. Encourage links at middle levels of PTTs, and knock on the doors of international standards groups and vendors with lists of suggestions and demands.

Some U.S. firms are doing these things, but too few, too slowly, and perhaps too late. What an opportunity is being missed!

Notes

1. This estimate turns up in a number of analyses of future employment scenarios. It generally includes workers in information- and communication-intensive industries, such as banking, as well as ones directly involved in the production of IT equipment and services. The best-known study of shifts in the composition of the U.S. work force relative to information technology is by Mark Porat, whose nine-volume study shows half the labor force now in information-related jobs (M. Porat, *The Information Economy*, U.S. Department of Commerce, Office of Telecommunications Special Publication, 77-12, Government Printing Office, 1977). Wilson Dizard provides a useful survey of the impact of IT on the economy in W.P. Dizard, Jr., *The Coming Information Age* (White Plains, NY: Longman, 1985).

2. These are 1989 figures, based on reports from the U.S. Department of Commerce. Obviously, relative labor costs are highly dependent on currency exchange rates that are often volatile.

3. Christopher A. Bartlett and Sumantra Ghoshal, *Managing Across Borders* (Boston: Harvard Business School Press, 1989).

4. This assessment is partly based on interviews and surveys carried out at Fordham Graduate School of Business in 1990, as part of a research study of information technology in the transnational firm. The interviews with senior business and IS managers in large firms indicated a general sense of uncertainty concerning transnational IT. The main uncertainties were: (1) Organizational design: what are the principles for evolving the transnational organization that should guide IT planning? (2) The global IT platform: how can the firm most effectively move toward a globally or regionally coordinated approach to IT planning? (3) Cultural integration and technology adoption: what are the main issues in managing not

just across borders but across cultures, the role of IT, and the impacts of cultural differences on IT planning? (4) Technology constraints, costs and opportunities: how can a transnational IT base be most effectively built, given the wide variations in cost and degrees of regulation across key international markets? See P.G.W. Keen, "Planning Globally: Information Technology Strategies in the Transnational Firm," in S. Palvia and S.P. Saraswat, eds., *Global Issues of Information Technology Management* (forthcoming).

5. The exhaustive keyword classification system for Information Systems research, developed by Barki, Rivard, and Talbot, does not even include international IT or use of IT by international firms as one of its 1,100 categories. The scheme was devised from a detailed review of IS research journals. H. Barki, S. Rivard, and J. Talbot, "An Information Systems Keyword Classification Scheme," *MIS Quarterly*, vol. 12, no. 2 (June 1988), pp. 299–322.

6. Richard A. Kuehn, "Consultants' Corner," *Business Communications Review* (May 1989), p. 88.

7. Much of the analysis in this section is based on a research study carried out by the International Center for Information Technologies on "How Cities Compete through Information Technology." Relevant reports are M.L. Manheim, J. Elam, and P.G.W. Keen, "Using Telecommunications to Gain Competitive Advantage: A Strategy for Cities," and D.W. Edwards, J. Elam, and R.O. Mason, "Securing an Urban Advantage: How U.S. Cities Use Information Technologies" (Washington, DC: ICIT Press, 1989).

8. The United Kingdom has about 40 percent of the European value-added network market and leads Europe in terms of local area network installations, low telecommunications costs, and electronic data interchange transactions.

9. See Manheim, Elam, and Keen, "Using Telecommunications to Gain Competitive Advantage," for a detailed review of the Netherlands strategy.

10. The *Financial Times* estimated in 1990 that consumers across the world were being overcharged for international calls by over $10 billion. U.S. calls abroad are on average 30 percent cheaper than those in the reverse direction. The U.S. deficit is over $2.5 billion. The international telephone companies use a cost-sharing formula that results, for example, in U.S. providers paying out 75 percent of the call charge to the PTT in the destination country. (See *Financial Times*, July 13, 1990.)

11. Today, a country wanting to build a private data communications network linking, say, the United Kingdom, Sweden, France, and Italy has to negotiate with the PTT in each country for both the originating and destination link (Italy to Sweden and Sweden to Italy), with long and complex schedules, tariffs, and installation. Seventeen European PTTs established an organization to provide one-stop shopping, including billing in a single currency. It never got off the ground. Internal squabbling among the PTTs, each anxious to protect its own turf and revenues, forced its dissolution.

12. "Hong Kong Designs Its First EDI Message," *Tradelink EDI Newsletter*, no. 4 (August–September 1989), pp. 1–2.

Chapter 4

Redesigning the Organization through Information Technology

Many, perhaps even most, large organizations are today approaching the limits of complexity. As business and government organizations grow larger and larger they are becoming less and less flexible and responsive. These increasing cumbersome and frustratingly bureaucratic organizations with often distant and impersonal management are ill-suited to deal with the dynamics of globalization, declining margins, time-dependent competition, and other destabilizers of the business status quo.

Few contemporary commentators view the large U.S. organization favorably. There is a sense of general malaise but little beyond home remedies to address it. Peters and Waterman, who asserted the continued viability of large firms in *In Search of Excellence* in the early 1980s, effectively recanted their view that such organizations could be creative, flexible, and responsive in their later works (see, for example, *Thriving on Chaos* and *The Renewal Factor*).

Notwithstanding the United States' reputation for having the strongest tradition of management education, consulting, theory, and development in the world, and a management elite that is almost obsessively introspective about culture, leadership, motivation, and communication, fewer and fewer observers today regard the structure and style of the traditional U.S. organization as an asset. Instead they see complexity and inflexibility—organizations that remain manageable only by adding complex procedures and layers of administration and staff that contribute little to the firm's mission and sometimes even impede it.

U.S. organizations need help. Nearly every leading authority on how to improve business performance in the next decade is calling for a drastic redesign of business organization. The reason that we *can* no longer have organization as usual is that we *do* no longer have business as usual.

Environmental Complexity

Business as usual has been rendered largely ineffectual by the growing complexity of the business environment. Several facets of that complexity, which were discussed at greater length in the opening chapters, are reviewed briefly here.

Globalization has extended lines of communication and coordination across time zones and locations, affecting breadth of markets, services, customer demands, and anticipation of competitive shifts. This hyperextension of activities is greatly straining the ability of traditional organizations to respond.

Time stresses resulting from geographic dispersion, shortened planning, development, and delivery cycles, and increased environmental volatility have drastically reduced acceptable reaction time. This is driving business not only to just-in-time inventory but to just-in-time everything: orders, scheduling, payments, manufacturing, distribution, and so on.

Change has become the norm, and unpredictability a basic reality of business. Discontinuities in the business environment affect the rules and invalidate old assumptions. Business must assume that the year 2000 will be at least as different from 1990 as 1990 was from 1980 and plan to be sufficiently flexible to accommodate a constant and rapid stream of change. Flexibility in the IT platform and extensive reach and range is an enabler of fast adaptation. Inflexible systems are a blocker.

Change may hit very close to home. Can any manager today state with reasonable certainty that his or her firm will not be involved in reorganization, relocation, acquisition, merger, or alliance? A firm's management must ensure that operations, communications, and management systems will be quickly adaptable to changes precipitated by such activities; can they do so without considering the firm's IT platform as an integral element in long-range planning? *No.*

Organizational Complexity

Most often, a firm's response to environmental complexity has been to increase organizational complexity. Management layers, procedures, and controls are added, with a concomitant increase in administrative overhead. Reliance on impersonal paper-based communication increases. The result is a host of organizational pathologies.

TENSIONS BETWEEN THE FIELD AND HEAD OFFICE GROW. Sales staff complain that they are not kept informed about products or have trouble locating the people who can answer their own or customers' questions. They view the corporate office as a remote "they" who do not understand the needs of the field.

LEADERSHIP IS DEPERSONALIZED. Senior executives rarely meet their own people other than for ceremonial visits. The CEO's business message and company news are weakly or belatedly disseminated via memo, rather than being communicated directly.

UNDERSTANDING IS FRAGMENTED. In a long chain of paperwork stretched over time, people, and geography, no single individual may have a complete picture of the system. However, many will have vested interests and independent priorities that create fiefdoms and unresponsive mini-bureaucracies.

PROJECT WORK AND TEAMWORK BECOME INEFFICIENT. When work is distributed across locations and lines of communication are stretched, coordination becomes expensive and complex, and is one reason why travel has become such a dominant and often debilitating feature of managerial life.

THERE IS GROWING SUBSERVIENCE TO DOCUMENTS. Large firms are often characterized by a virtual ritualization of documents: travel expense claims, purchase order authorizations, requisitions, and so forth. These documents can easily take over the work process they are intended to support, controlling rather than being controlled by the people who must rely on them.

MIDDLE MANAGERS FACE A FUNDAMENTAL DILEMMA. Corporate America, facing up to its eroding international competitiveness, is aggressively looking for ways to change management processes and reduce the labor component of products and services. An executive in one company that is transforming the basics of its operations remarked that "costs walk and they usually wear a suit"; organizations are taking every opportunity to walk those costs out the door. At the same time that middle managers are coming under increasingly heavy budget pressures, they are being asked to adapt rapidly to new rules, erosion of the old hierarchy, and loss of many of their perquisites and authority. Old assumptions of continued job security, if not lifetime employment, are gone. The typical 50-year-old manager in a *Fortune* 1000 firm has no reason to believe he or she will still be in the company at 60. The dilemma for middle managers is that they have the most invested in the status quo yet are the cornerstones in the implementation of changes to undo the status quo.

EXPERIENCE ASSUMES A NEGATIVE VALUE. In a context of constant change, experience often becomes a liability rather than an asset. The value of experience generally rests on the status quo—the less things change, the more experience is worth. The more things change, the more people have to invest in continuous self-education.

IT Countermeasures

The pathologies just discussed are well recognized and many firms are struggling to treat them, with varying degrees of success. Most, however, have overlooked opportunities to use IT to attack their root causes. These tools include videoconferencing, electronic data interchange, business television, laptop computers, and CD-ROM (compact disk-read only memory).

In most organizations IT has *added*, not reduced, complexity. Organizational complexity—increased layers of management and staff, administrative overhead, formal control and reporting systems, and the substitution of impersonal paper for people in communication—has often resulted from the introduction of clerical automation and management information systems. The volumes of computerized reports and controls that the IS function have created have replied to complexity with complexity.

Most of the information technology that could be used to reduce organizational complexity has been available for five to ten years. That it hasn't been so used reflects a narrow perspective on the part of many Information Services units, which have traditionally viewed "information" as computer data bases. Most of the IS effort through the 1970s went into transaction processing and data base management, and the mainstream of the technology was built to support this. Though the office technology of the 1980s provided new tools that focused on text and documents, the mainstream skill base in IS continued to rest on managing numeric data. Much of the complexity of organizations comes from documents, and much of the opportunity for creating organizational advantage through organizational simplicity rests on shifting attention from data to documents. The real payoffs from using IT rest far more on managing documents electronically and facilitating fast, natural, and simple communication than on managing data. Figure 4-1 links organizational complexity to environmental complexity, lists the derivative organizational pathologies, and identifies five ways IT can be used to reduce organizational complexity by:

1) Targeting organizational simplicity of work procedures and co-ordination as a source of organizational advantage;
2) Designing structure- and location-independent organizations;
3) Facilitating the collaborative organization;
4) Repersonalizing management; and
5) Making it easier to communicate than not to do so.

The use of IT to reduce organizational complexity is limited mainly by imagination, creativity, and business management commitment, not by technology or cost. Among the most practical and badly needed areas of application of IT: increase direct, flexible access to people and reduce the need for information intermediaries; provide simple access to information, simply organized; focus on people's needs for document-based information; reduce paper document flows and barriers to tracking, locating, and controlling document-based work activities; and enable a committed drive from the top of the business to cut superfluous layers of staff and management.

The last point is not peripheral. Organizations do not delayer themselves, nor do well-entrenched procedures relax and reform themselves in response to the logic of simplification. All too often, social inertia dampens efforts to change. Jobholders, however well intentioned and committed to a firm, have an investment in the status quo and a set of skills and experience built on and often locked into entrenched tradition and equilibrium. In the firms that have successfully used IT to break out of the status quo, the business leadership has added its weight to the push.

FIGURE 4-1 **Organizational Complexity: Causes, Consequences, and Solutions through IT**

ENVIRONMENTAL COMPLEXITY

- Globalization
- Hyperextension of operations
- Time stresses
- Discontinuities (political, business, economic, social)
- Business restructuring (reorganization, relocation, acquisition, merger, downsizing)

Gives rise to

ORGANIZATIONAL COMPLEXITY

- More managerial layers
- Elaboration of procedures and controls
- Administrative overhead
- Reliance on communication by paper and reporting systems

Attack

Gives rise to

IT COUNTERMEASURES

- Re-create organizational simplicity
- Design structure- and location-independent organizations
- Facilitate the collaborative organization
- Make it easier to communicate than not to do so
- Repersonalize management

ORGANIZATIONAL PATHOLOGIES

- Field/HQ tensions
- Depersonalization of management
- Fragmented understanding
- Inefficient project work and teamwork
- Subservience to documents
- Middle management dilemmas
- Negative value of experience

Organizational Simplicity

Complexity erodes flexibility, responsiveness, and morale; simplicity increases them. U.S. organizations are reaching the limits of complexity but not size.

IT mechanisms that can be used to reduce organizational complexity include videoconferencing, which provides simple face-to-face communications, "team technologies" such as organizationwide (and hence platform-dependent) uses of electronic mail and "groupware," and every application of IT that eliminates delays, administrative intermediaries, and redundant steps in transactions and that improves access to information. Electronic data interchange and customer-to-supplier links are examples.

Telecommunications is the basis for much of organizational simplification. In the late 1970s McKesson cut the size of its purchasing department from 700 to 15 people by allowing pharmacists to place orders directly into its computers and by linking its own systems directly to those of its suppliers. This organizational advantage translated into a distinct competitive advantage throughout the 1980s.

Bureaucracy and bodies dampen innovation, communication, and service; fewer layers, fewer people, fewer administrative steps, and fewer sources of bureaucracy, error, paper, and procedures add up to an organizational advantage that at the very least makes a company a healthier environment and probably contributes in the long term to competitive advantage.

By enabling information to be moved quickly to where it is needed, telecommunications reduces "information float," the time gap between the occurrence of an event and the information about the event becoming available. Like check float, which represents unavailable funds, information float represents wasted time. Just-in-time business needs just-in-time information.

Allowing managers and staff to access information directly instead of having to rely on intermediaries cuts out two levels of organization. Wells Fargo's personnel department, by providing personal computer access to on-line personnel data, accomplished such a delayering with an increase in productivity. Similarly, a major British firm found that the use of laptop computers eliminated most of its branch meetings for field sales personnel, subsequently realized that the entire branch structure was obsolete, and cut two layers of management.

Videoconferencing provides a cautionary example of how the traditional perspective on IT has limited exploitation of proven tools. Videoconferencing has been in use for more than twenty years to support project management, education programs, and ad hoc team efforts. But as so often happens with IT, early expectations have not been met. Videoconferencing remains an organizational eccentricity in most firms. In firms that have them, video rooms are often used for only an hour a day; electronic meetings have not become part of the structure of everyday business life.

There are at least four possible explanations for the underutilization of videoconferencing: the cost of the technology; lack of an organizational home; wariness about using an unfamiliar medium; and inappropriate business justification.

Before 1990, the cost of equipment, telecommunications links, and special rooms was sufficiently great to discourage experimentation with video technology, and the direct cost savings from reduced travel was too small and uncertain to permit easy business justification. Today, an in-house video room can be built for less than $50,000, and satellite and fiber optics have reduced transmission costs by a factor of ten or more since 1985. In addition, there are many commercial providers of conferencing facilities; US Sprint and several hotel chains rent video rooms for as short a period as half an hour. British Telecom provides similar services connecting cities in Europe, Australasia, Asia, and North America. A reliable rule of thumb is that the cost of a two-hour videoconference using commercial services is the same as first-class air fare plus hotel and standard per diems.

Because videoconferencing is neither a standard responsibility of the IS department nor a natural area of activity for any major staff, line, or administrative group, it requires a fairly senior champion. Most often today the champion is a manager who is motivated by more than a desire to cut travel bills, who sees videoconferencing as a means to organizational coordination of core activities, repersonalizing leadership, and gaining organizational flexibility and responsiveness.

Unfamiliarity with the medium leads many managers to react negatively, citing numerous examples of meetings in which shared physical presence was vital. These same managers often will have discussed important and delicate issues with a colleague on the telephone. In practice, few people find videoconferences stiff, unnatural, or lacking in social nuance and a sense of personal contact. Properly designed and operated, the video screen becomes an extension of local surroundings.

Trying to justify videoconferencing as a substitute for travel is a mistake; in practice, it has little more impact on travel than the telephone and even when it does reduce trips, the direct savings rarely pay for the needed capital investment. One of the biggest challenges for proponents of any value-added use of IT—to make a convincing economic case—is still largely unmet. If videoconferencing does not save money, how should it be justified? There is no stock answer to this question, for videoconferencing or for other office technologies, executive information systems, and decision support tools. All aim at improving "effectiveness." They can be convincingly justified, but only by being clear about the nature of the benefits and relating them back to core business and organizational drivers. "Convincingly" does not necessarily mean "quantitatively."

Despite blockages of the widespread diffusion of videoconferencing, it is growing in use rapidly in more and more firms. Effective applications include coordination of buying operations, support of continuing education, and crisis management.

J.C. Penney has demonstrated that in-house business TV can significantly improve core operations. Its buyers "meet" weekly by video to place orders instead of traveling to the head office several times a year. Because the goods can be seen, there is no need to wait for samples, and orders can be placed directly and immediately. The result has been lowered costs for inventory and samples, shortened order cycles, more local discretion for store-level buyers, and the elimination of inventory markdown sales. New buyers are trained in videoconference participation and presentation. In 1989 Penney extended the reach of its system to the Far East, sending still-image video transmissions of samples, designs, and specifications to its private-label suppliers. It has transformed a complex procurement process that was routinely marked for a 14-day turnaround for even simple inquiries and orders; today they take just hours, with marked improvements in quality control, since Penney and its suppliers can add more frequent verification and get fast answers to queries.

Hewlett-Packard routinely provides for its engineers weekly education programs taught by faculty from Stanford University, using the electronic links to minimize travel and disruption. Students are spared the drive to the university; teachers avoid traveling to company headquarters. Hewlett-Packard's chairman, John Young, sees business TV as a major support of the firm's business strategy. He states that "As time-to-market becomes a critical competitive advantage, 'just-in-time' communications and training becomes the key to business success." Domino Pizza's president, Don Vlcek, similarly comments: "We're running a $3 billion company made up of 5,000 little stores. With only 60 people at headquarters we've *got* to use technology to inform and train everybody."[1]

Nearly every notable commentator on U.S. business has stressed the need for massive and sustained education—remedial training as well as professional and management development, technical skills updating, and product and marketing training. Motorola provides high-school-level training in basic literacy and numerical skills for almost half the staff it recruits. Louis Gerstner, president of American Express, writes that in the next five years the company will hire a minimum of 75,000 people. "That's the good news. The bad news is that we may not be able to find them." American Express is having to spend $10 million a year teaching employees skills they did not learn in high school.[2] A number of firms in the high-tech field, where the half-life of knowledge is shortening as fast as the technology is accelerating, calculate that keeping up to date in a specialized field demands half a day of education per week—10 percent of an employee's salaried time.

Education is expensive and first-rate teachers are scarce, especially in the IT field. Releasing people to attend a three-day course can sometimes be justified but is disruptive and inefficient on a regular basis, and bringing the best, and busiest, teachers to a firm or to a nearby conference center is often not practical. It is much more efficient and eminently practical to bring the students and educators together electronically.

As it does for travel, videoconferencing augments rather than substitutes for traditional face-to-face education. It enables firms to expand their education delivery base quickly and effectively. The National Technological University, a consortium of colleges across the United States, has shown the applicability of videoconferenced education in broadening curricula and providing materials its members could not have developed on their own.

In the early 1980s, the pilot use of a videoconferencing capability linking key British and German engineering and production locations of Ford of Europe did not justify full-scale implementation, viewed from the perspective of its being a travel substitute. What convinced top management to approve its expansion was its potential for crisis management and its capacity to bring together ad hoc teams on very short notice. Ford's effectiveness as a firm depends heavily on cutting "time to launch"—the roughly five-year time frame for bringing a new vehicle from design to market. Every day's delay costs hundreds of thousands of dollars. In the final pre-launch phase, in which a car is tested to destruction, any breakdown is a "job stopper." Teams of British engineers are rushed to Germany to review the problem with their German colleagues. Or they are rushed into the video studio, cutting several days out of the job stopper.

Location- and Structure-Independent Organizations

Beyond simplifying the organization—by delayering it, tying it more directly to customers and suppliers, and bringing field and headquarters staff, educators and students, and crisis management teams together electronically—IT can enable entirely new, productive forms of organization. Used in this way, IT no longer merely supports the business; it becomes a force for organizational invention as well.

Historically, firms have brought people to work and relied heavily on organizational structure as the basis for operations and strategy. Today, firms can bring work to the people and begin to contemplate the design of their organization the way they design products, from first principles, while being less and less constrained by limits of time and place. A firm's telecommunications network can enable it to locate its customer service staff in Heathrow, Florida, or Millinocket, Maine, even though its headquarters, along with all relevant records, is in New York. Just as a viewer who plugs a TV into the wall and switches it on does not know or care where the electricity "is" nor how CBS or HBO is transmitting the signal to the TV, the customer service workstation user is not concerned that the information is "in" New York. Similarly, the fact that India severely restricts the availability of computer hardware has little impact on Citibank's software development in Bombay. Satellite links effectively put the needed hardware, which is physically located in New York, "in" India.

Telecommunications enables a firm to geographically link separate organizational units, relying on electronic mail, facsimile, videoconferencing, and workstation access to shared information resources to coordinate across time zones. This is now common practice. It is difficult to assess the nature and impact of electronic communication on the basics of organizing and coordinating. The tools are fragmented and often used only by some units in a firm; aggressive overselling and hyping by proponents of the "office-of-the-future-is-now" view and by vendors has engendered a customer backlash. It often takes many years for social change to be fully observable and assessable.

What is not routine yet, but has fairly immediate and significant impacts on organizational work, is the use of telecommunications for location-independent operations. Recent examples include Bechtel's plans to offshore people in the same way that companies offshore manufacturing. Bechtel, which manages construction and engineering projects across the world from its main offices in San Francisco, has traditionally moved its people to the work: Middle Eastern oil fields, Latin American dams, Boston bridges, and African power plants. A typical Bechtel manager will have changed location every one and a half years. The president of one of Bechtel's business units has moved 27 times. Shortage of available technical staff—a growing problem for most U.S. firms—and shifts in the willingness of employees and their families to live this way make such mobility increasingly impractical and unattractive.

Bechtel is now moving work to its people. Very small aperture terminals (VSATs) that send and receive information by satellite, electronic data interchange, and distributed computing will enable the company to assemble, for example, a team of engineers in Ireland that is electronically linked to the project location and to San Francisco for access to engineering information.

Several transnational firms are creating electronic centers of expertise for tax functions, finance, engineering, and back-office operations. When one petrochemical firm, which has all its worldwide tax specialists in London, opened a branch office in China, it did not need to hire an expert there or in Hong Kong, since its London expertise was easily accessible to planning staff in both New York and the Chinese branch. Similarly, the firm's worldwide center for engineering in New Jersey provides links to and from the computers, information, and staff at its other international and domestic locations.

Large transnational firms are clearly moving to a mode of coordination that ends the old dichotomy between centralization and decentralization and creates centralization-with-decentralization and substitutes coordination and collaboration for control. Location will no longer determine planning, control, reporting, function, and communication, making the firm's telecommunications resources the real definer of "structure."

Two of the most basic assumptions of modern organization theory and practice are (1) that the organization is its formal structure, and (2) that strategy and structure are intimately interrelated. "The formal structure of the organization," write Miles and Snow, "is the single most important key to its functioning. . . . Historically, strategy and structure have evolved together."[3] This is just one of many assertions of the dominance of these assumptions in organizational research and much of management practice.

IT increasingly invalidates these assumptions. One example of an

IT-based form of organization that does not readily fit traditional assumptions about structure and strategy is the "metabusiness," a quasi-firm created through electronic linkages between organizations that so tightly couple participants' operations that it is impossible to say where the boundary of one firm ends and the boundary of another begins. Metabusinesses are structure-independent; participants needn't have the same structure and can change their own structure without affecting the other participants.[4]

An early and obvious example of a metabusiness is General Motors' supply system, which effectively brings its suppliers' operations into its internal organization. GM has substantially eliminated boundaries between itself and its suppliers, though legally they remain separate entities. GM requires suppliers' computers to link to its own in order to eliminate paper transactions such as purchasing and invoicing, reducing the need for human administrative intermediaries.

Such tight interfirm linkages generate entirely new dilemmas of management. For example, in the face of problems or queries, one supplier asks: "Who does what at what time with what authority? . . . We now run the factory for GM. GM basically tells us what to do." Buick Epick, an electronic linkage that handles warranty, ordering, and customer service transactions between GM and its Buick dealers, alters legal boundaries and responsibility between dealer and supplier. Who, for example, is liable for and owns transactions and the right to use the resulting data and knowledge? What are the boundaries of the resulting "organization"?

The extreme of the cross-everything organization is Galoob Toys, a $60 million company with just 115 employees. Galoob farms out manufacturing and packaging to a dozen contractors, uses outside distributors without ever taking delivery from the manufacturers, and sells its accounts receivables to a factoring company. Its business is one of relationships.

Such "dynamic networks" (a term invented by Ray Miles) are becoming more and more common as U.S. firms try to be rid of overhead functions, exploit offshore manufacturing, and maintain flexibility. "What you'll have," says Miles, "is a switchboard instead of a corporation."[5]

This is very much the case with the world's major airline reservation systems. Consider the travel agent in Princeton, New Jersey, who books a business passenger from New York to London and then on to Paris and Amsterdam and back to New York after a stopover in Miami. The agent interacts with what looks like a single travel reservation service that handles bookings for perhaps four airlines, five hotels, and two car rental firms. The travel agent has no idea where the providers of the service are. If he or she is using Covia's Apollo system, the computer is in Denver and may link to American Airlines' system in Tulsa or British Airways' London system. The travel agent has no reason to know or care.

Such electronic relationships are structure-independent. Acquisitions are not; they require one or both parties to make major changes in management, formal procedures, and so forth. An agent in a metabusiness may be vulnerable to control by other agents, but each partner can operate without changing its structure to fit the other's. This makes the constant focus on structure in organizational literature increasingly unhelpful and unrealistic.

SWIFT (Society for Worldwide International Funds Transfers), a long-established worldwide electronic utility shared by a growing consortium of banks, raises the question of what exactly an organization is in a context of fully coupled, shared electronic business operations. SWIFT changes many of the basic dynamics of its members' businesses. For example, it requires competitors to cooperate. Walter Wriston provides an elegant summary of modern electronic banking: "Cooperate in the morning. Compete in the afternoon." By cooperating to create SWIFT, banks can compete against each other in selling electronic cash management systems.

SWIFT also establishes close structure-independent linkages between firms—Citibank, for example, can interlink to Chase and Deutschebank without making any structural accommodation—and makes adoption of a computer or telecommunications standard or protocol no longer a technical issue but a basic determinant of the available range of business options. A bank that does not adopt the SWIFT format for funds transfers is locked out of a wide range of electronic services.

Traditional organizational theory and the language it has created are increasingly inadequate to describe the metabusinesses that are becoming the norm in industry after industry. Such businesses are making many of our most established assumptions about organization structure and its relation to business strategy obsolete.

The Collaborative Organization

The business team, rather than the functionally defined hierarchy or the dotted-line or matrix-management variants that show up on the organization chart, is increasingly being seen as the real unit of organizing. To some extent, teams have replaced "excellence," "culture," and the like as the latest fad.[6]

But glibness about teams should not distract us from the reality that complex environmental, societal, and economic changes are pushing organizations toward new forms of collaboration. These pressures are destabilizing established routines and structures and placing a premium on mechanisms for supporting rapid adjustment to new and unpredictable situations. This inevitably involves collaboration across previously separate boundaries—between functional areas, locations, companies, and countries. Team-based structures and processes emerge naturally in this context.

Change as the norm has resulted in a shift in emphasis from organizing by division of labor to organizing by division of knowledge. This is the real force behind the emergence of the team-centered organization as a necessity for adaptation. With division of labor, which depends on relatively predictable tasks and demands for knowledge, experience is an asset. Almost by definition, the highest up the hierarchy, the more knowledgeable the job-holder. There is a clear link, in theory if not in practice, between authority and expertise.

The term "division of knowledge" captures an obvious reality of work in an era of rapid change and uncertainty. Tasks are no longer predictable and experience may no longer be valuable. New inputs of knowledge are needed to define tasks, and multiple skills and experience are needed to complete them.

Division of knowledge has long been the basis of much work by Information Systems development teams, R&D units, and task forces. More and more areas of work are moving into the same zones of uncertainty and change that have characterized the specialized technology and research functions of the firm. In light of this, the shift toward business teams is less an innovation than a catching up. Business is finally recognizing that division of labor is increasingly ineffective as the basis for an organization in an environment of constant rather than occasional change and of shifting functions to the proven project- and team-based processes long employed by the research and technical functions.

Teamwork is relational; the quality of performance rests on the quality of interactions, communication, and coordination among team members. Management control is replaced by management coordination of the work of others who may know more than the manager, and decision making occurs in the team rather than in the hierarchy.

The basis for effective teamwork is collaboration, "shared creation" according to Michael Schrage, the author of an insightful book that deals with the contribution IT tools can make to collaboration.[7] Collaboration is a joint commitment to a target output, with team members sharing authority and responsibility as needed, at different stages and for different tasks. Many companies talk of the need for teams to be "self-managing." Situational authority is a key element in collaboration, an obvious example being a theater or ballet company; in the interest of achieving the best performance, stars in effect loan authority to the stage manager, conductor, and director.

Recent studies have highlighted the sequence of adaptation to the globalization of business that leads to the collaborative team becoming more than just an occasional task force and temporary mechanism. Unilever's experience is typical. As it attempted to bring a global perspective to its core functions and operations, the firm discovered that decentralized country-by-country responsibility and authority, the mark of the traditional multinational, could not handle the interdependencies among previously separate business units. Management recognized the need for joint planning to handle these new demands, but the old structures impeded action. Unilever then used its central staff to "coordinate," which merely added layers of management and disagreement. When corporate management's fallback to the "pretty please" approach elicited largely feigned compliance, Unilever began to build real incentives and mechanisms for collaboration, an effort that took years to accomplish and required changing an entire management style.

Bartlett and Ghoshal found this pattern of adaptation to be typical in the nine firms that provided the basis for their book, *Managing Across Borders:*

> In company after company, the most difficult challenge was to develop the new elements of the multidimensional organization. . . . They built and managed interdependencies [and moved] from control to coordination and cooperation. . . . The most successful did so not by creating new units, but by changing the basis of the relationships among product, functional and geographic units.[8]

Bartlett and Ghoshal and other commentators provide many other examples of how this is achieved and of the benefits that accrue. For example, Ford, which credits the development of its magnificently successful Taurus line of cars to the teams of stylists, engineers, and production personnel who created a new mode of working together, has announced that Team Taurus will serve as the model for all future development efforts.

Many successful and adaptive firms use lateral job assignments, cross-functional and cross-geographic forums and meetings, and training as vehicles for breaking down fixed boundaries. Ikujiro Nonaka, a Japanese professor, points out that Japanese firms move people across functions and locations on a regular basis. He cites Honda as an example of how cross-fertilization and broadening of knowledge builds an ongoing organizational asset.[9] Studies of successful Japanese firms suggest that they are more effective team-builders than most U.S. firms, in part because they maintain flexibility in roles and empower middle managers who are good collaborators and communicators to become coordinators of team-led work.

All this is a fairly recent development; the Japanese firm of the 1950s looked much like the stereotypical German or British firm—rigidly structured, with a wide gap in status and pay between manager and worker, formalized functional divisions of labor, and decision making and planning the prerogative of top management.[10] Leading Japanese firms seem to have discovered far earlier than most U.S. and European firms the fairly simple reality that change and complexity in work require collaboration across flexible teams built on division of knowledge.

Leading non-Japanese firms are moving rapidly in the same direction. The chairman of Digital Equipment Corporation, Ken Olsen, believes that the ability to coalesce teams electronically is one of the most important functions of telecommunications and computing. DEC, like General Electric, Citibank, American Express, and many other companies, electronically links all of its offices. Most of the firms known to emphasize teams have a worldwide telecommunications network capability. Thus a major priority for any manager trying to make the business team a driving force for adaptation must be to ensure that IT investments are targeted toward the same end.

That this is not yet self-evident is apparent in the history of video-conferencing. If enough IS people and business managers were thinking in terms of technology support for teams, a first-rate video-conferencing infrastructure would be mandatory. Similarly, every employee's electronic mail address would be on his or her business card. In general, efforts to use IT to support collaboration, communication, and coordination have been fragmented. That is sad news. Despite the abundance of IT tools that might be developed into a coherent organizationwide base for team support, many companies have only limited and often ad hoc applications in place at the very time that they are making major efforts to improve communication and teamwork.

One of the main reasons for the sluggish adoption of such potentially valuable IT tools as electronic mail, videoconferencing, groupware (the fashionable term for personal computer software designed to support teams rather than individuals), and decision rooms has been the lack of a common delivery base and of consistent common technical standards. Many electronic mail systems cannot interact with one another, and few firms have the corporate IT platform needed to use different technologies in a coherent fashion.

Team technologies are as effective as their degree of "reach"—the locations to which they can connect—and "range"—the variety of information they can share. These two dimensions define the IT platform. Generally, business applications of IT can be fairly effective, at least in the short term, even with limited reach and range. But team technologies are significantly hampered by the lack of a companywide platform. If we accept that such uses of IT are vital to the firm's organizational strategy, the positioning of the platform becomes an important element of business, not just IT, planning.

Repersonalizing Management

The need for communication and leadership are among the most widely accepted axioms of modern management. How can leaders motivate by memo? How can they really communicate across an organization spanning three to twelve time zones? How can they communicate to employees their own values and personality rather than just abstract mission statements and reports?

The number of companies using business TV and videoconferencing as a vehicle for managers to communicate directly with the wider organization, virtually face-to-face instead of by memo and corporate communications' glossy publications, is growing rapidly. One of these firms is Federal Express, a company that built in just over a decade (from scratch) a powerful business based on a clear vision, strong commitment to treating people well, strong expectations of their performance, and information technology. Business TV is an important part of this mix. Federal Express promises its staff that they will hear news about the company from the company before they hear it from the press. When the firm arranged to buy Flying Tiger, the news was broadcast through the in-house business TV system just minutes after the news was released publicly. Fred

Smith, the founder and CEO of Federal Express, comments that "It would be impossible to place a dollar value on that, but I have no doubt that it left a powerful and lasting impression on the thousands of employees who saw that broadcast." Without the electronic communication base, the alternatives would have been (1) send out a memo and let most people read the news in *The Wall Street Journal;* (2) make a few phone calls and leave matters to the grapevine and rumor mill; or (3) stick notices up on bulletin boards.[11]

One of the truisms of today's management is the importance of communication. What does this really mean? Many studies have found that the messages in corporate communication programs often do not get through and that corporate management is seen as remote. Employees surveyed consistently report that the communicator they most prefer is their immediate boss, the one they see most days and with whom communication is two-way and face to face.

Tom Peters, long a consistent advocate of taking communication seriously, talks about effective firms in the emerging "life without hierarchy" as marked by "rich and dense connections" and much improved "odds of connection."[12] He argues that increased competition paradoxically demands increased cooperation.

That is exactly what video technology provides: the richness and denseness of face-to-face and (often) two-way communication and far better odds of connection. The Federal Express example is a small and undramatic one; it provides a base for new modes of ad hoc, occasional, and routine communicating. This is no big event; it is what communication surely means—no big event but continued connection.

Building the "Relational" Organization

Information technology affords the opportunity to build what might be termed the "relational" organization: an organization defined not by fixed structures but by ease of relationships. Instead of focusing on organization structure, business today needs to look at the mechanisms that make communication simple, flexible, and natural. IT makes practical many of the visions of management thinkers: Peter Drucker's network organization, Rosabeth Kanter's "dancing elephants," Stanley Davis's "future perfect," and Tom Peter's life without hierarchy.[13]

What IT platform is needed to provide these mechanisms? It is no criticism of the visionary gurus who stress the need for new forms of communication and coordination to say that they simply assume that the needed IT resources will be in place. Peters calls this a competitive must. But who will explain what technology judgments and decisions will be required to develop an adequate corporate IT platform? The vision demands the technology, but the technology cannot substitute for the vision. The corporate IT platform will determine the enabling IT mechanisms, and these will become the effective organization structure.

Just as it is becoming impossible to define a business plan that is not strongly dependent on IT, it is irrelevant to discuss a communication strategy in a large, geographically dispersed organization that does not include IT, particularly a platform for team technologies and a videoconferencing infrastructure. One of the major inadequacies of incompatible IT resources is the inability to communicate across functional and locational boundaries.

The relational view challenges much established organizational theory and practice, particularly the assumption that the organization is defined by its formal structure and that strategy and structure must move together. Ending structure dependence changes the nature of potential strategy.

Consider the following description of a small company, a diemaker that has "organized itself around an electronic network." The authors intended that it be read in terms of competitive advantage; read it instead in terms of organizational advantage, of how relationships are dramatically changed. It is a relational organization; successes and breakdowns reflect the quality of relationships, and the network is the basis for collaboration, cooperation, teams, partnerships, and joint ventures.

> Designers at Hitachi sketch a new part and send it by fax to the diemaker. Die engineers review the sketch and, using computer-aided design (CAD) systems, generate the specifications for a new die in a matter of hours. The company then decides whether to make the die itself or subcontract to one of the suppliers—all of whose skills, current capacity, and work-in-progress have been logged. As often as not, it chooses a supplier and sends the specifications by fax.

> The supplier, using advanced numerical control tools, makes the die, also in a matter of hours. It is not uncommon for Hitachi to get the die for some parts back in a day.[14]

Today, this firm is a prototype. It will surely be the norm by the mid-1990s.

The management principles for exploiting IT to organizational advantage and as a powerful force in organizational redesign require a basic rethinking of old assumptions—especially concerning the links between strategy and structure—at the top of the firm. A management act of will is required. No mid-level business unit or IS manager can create and implement policies and plans that redesign the way an entire organization works, communicates, and coordinates. The requisite principles are essentially the scrupulously thought-through application of the IT countermeasures elucidated above.

Notes

1. These quotations are extracted from J.M. Douglas and D.P. Brush, "The CEO and Business Television," *Business TV* (September–October 1989), pp. 41–44.

2. L.V. Gerstner, "One Nation, Underqualified," *Washington Post*, May 16, 1990.

3. R. Miles and C.C. Snow, "Designing Strategic Human Resource Systems," *Organizational Dynamics* (AMA, 1980), pp. 69–82.

4. For a more detailed discussion of metabusinesses, see P.G.W. Keen, "Rethinking Organizational Design: A Relational Framework" (Washington, DC: ICIT Press, 1989).

5. Quoted in J.J. Wilson and S. Dobrzynski, "And Now, the Post-Industrial Organization," *BusinessWeek*, March 3, 1986, pp. 64–71.

6. For a comprehensive analysis of team technologies, see R.R. Johansen, *Groupware: Computer Systems for Business Teams* (New York: Free Press, 1988).

7. Michael Schrage, *Shared Minds* (New York: Random House, 1990).

8. Christopher A. Bartlett and Sumantra Ghoshal, *Managing Across Borders* (Boston: Harvard Business School Press, 1989).

9. I. Nonaka, "Toward Middle-Up-Down Management: Accelerating Information Creation," *Sloan Management Review* (Spring 1988), pp. 9–18.

10. See R. Clark, *The Japanese Company* (New Haven: Yale University Press, 1979).

11. Douglas and Brush, "The CEO and Business Television," pp. 43–44.

12. Thomas J. Peters, *Thriving on Chaos: Handbook for a Management Revolution* (New York: Alfred A. Knopf, 1987).

13. P. Drucker, "The Coming of the New Organization," *Harvard Business Review* (January–February 1988), pp. 45–53; Rosabeth Kanter, *When Elephants Learn to Dance: Managing the Challenges of Strategy, Management, and Careers in the 1990s* (New York: Simon & Schuster, 1989); S.M. Davis, *Future Perfect* (Reading, MA: Addison-Wesley, 1989); Thomas J. Peters, *Thriving on Chaos: Handbook for a Management Revolution* (New York: Alfred A. Knopf, 1987).

14. K.B. Clark, "What Strategy Can Do for Technology," *Harvard Business Review* (November–December 1989), pp. 94–98.

Chapter 5

Redeploying Human Capital

Business needs to learn to treat people like machines. It accords the machinery of IT—the hardware, software, and other components—care, long-term planning, and commitment. Every large company that invests heavily in IT spends substantial sums of money maintaining its IT equipment and has a hardware plan that typically assesses technology changes and vendor offerings three to five years hence. Much rarer is the firm that acknowledges the importance of education, which is the equivalent of maintenance for people, and has a formal organizational plan that looks ahead in detail at job, career, and skill changes and needs. This situation is surprising in light of the fact that (1) the human element is the critical facilitator or bottleneck to effective use of IT, especially as the technology becomes more cost-effective and easier to install, and (2) IT can quickly and almost completely erode the value of their experience, create demands for totally unfamiliar skills, and stop careers dead.

The experiences of leading U.S. firms in fast-changing fields such as pharmaceutical research, high technology, and advanced engineering suggest that when workstations become a core feature on the office, store, or factory landscape, people need to spend 10 percent of their time on education. This amounts to half a day a week. It is not the technology itself that causes experience to lose its value—computers have been around for close to forty years, personal computers and word processors for ten, and telecommunications is hardly new—but rather the use of IT to rethink, rather than simply automate, the status quo. In the early 1980s, for example, major banks began to shift much of their corporate service away from loan-based products toward "relationship" banking, built around fee-based services such as electronic cash management, automated letters of credit, and electronic lockboxes, that depended heavily on IT. Some corporate bankers who found themselves needing to sell products they did not understand, built on a technology they neither valued nor used, resisted or ignored these trends. Their implicit assumption was that credit was difficult to learn, hence experience was valuable. That may still be, but credit experience is no longer enough. Marketing electronic products demands both a detailed understanding of clients' operations and internal systems and a new style of consulting.

Many of the experienced telecommunications managers called upon to support these new services had learned their trade in a regulated environment of voice communications and telex, experience that was becoming less and less valuable in a world of deregulation, data communications, local area networks, EDI, image technology, and a flood of new applications, tools, and technologies. These managers needed both to update their knowledge of the technology itself and to learn to think in business terms about its impacts and opportunities.

Flexibility and adaptability are the new watchwords for organizations and for people. Many of today's jobs will disappear and many of today's people lack the skills needed to do tomorrow's jobs. Firms that do not attend to their people's careers as well as their jobs may well have to replace much of their work force. Their human capital will be a depreciated resource, one they may find harder to replenish than they thought.

People and Careers in the Information Age

What is my career, as opposed to my job? What is likely to happen to my job? What IT skills and knowledge might I need and why? Can I determine whether specific experience will become more valuable over time or become a commodity? What education must I acquire to ensure that my career, skills, and experience are not quickly depreciated? These questions are as relevant for senior business managers as they are for IT users, and for IT professionals as well as for business people.

There is a major difference between a job and a career. People own their careers; their jobs are owned by their employers. Gone, probably forever, are the days when most people entered a specific job function in a large organization and moved up a career ladder, aware of the general patterns and timing of promotions and the knowledge and skills needed to advance. Those were the days of the career "path," when careers were the sum of a set of jobs.

When information technology is used to redefine the very basics of business, it brings radical change to work and thus to jobs and careers. Increasingly, the pace and effectiveness of business innovation through IT depend on people rather than technology. There is a wealth of literature on "resistance to change," the impacts of IT on workers' job satisfaction, techniques for creating user "involvement," and many other aspects of the human side of IT.[1] Technology capital and business capital depend on human capital—"capital" in the sense of skill base, education, relevant experience, and career development.

It is imperative that firms have strategies for managing the human as well as the technical side of IT. This means rethinking many of the standard assumptions and management practices concerning jobs, careers, career development, human resource management, and education.

The axioms for work in the coming decade are likely to include the following:

Continued education is essential for employees and employers alike.

No one is unaffected by or able to ignore information technology.

Change is the norm.

Work is highly interdependent, involving business teams, cross-functional communication, and lateral moves into other areas.

There are no standard career paths.

To this emerging reality must be added a sustained management recession, a thinning of the middle management ranks of the *Fortune* 1000 and its European equivalents (organizations of the 1990s will typically be 20 percent smaller and two management layers lighter), and an overall decline in the number of jobs (estimates of the impact of IT in most industrialized countries call for reductions of 30–40 percent in office jobs, clerical and administrative positions, and many areas of service industries by the end of the century). This will be offset to some extent by a shortage of younger employees.

One major U.S. hotel chain views as its competition in the 1990s not other firms in its industry but McDonald's and the U.S. Army. All will be trying to recruit eighteen-year-olds. The hotel calculates that it will be able to hire only two people to fill three jobs and views IT as the only way to compensate, by putting technology behind staff in what used to be largely manual functions.

The picture we paint for the 1990s is thus one of increasing displacement of mid-level personnel; a shortage of junior employees; a demand for new skills in many of the new jobs created; old jobs redefined by information technology; and education as a basic requirement for career development. It is education that will help people adjust to new situations, learn how to rethink aspects of work altered by IT, and exploit the opportunities afforded by new technologies.

Career Trajectories

What are people to do about career planning in the vastly different work landscape of the 1990s? The framework in Figure 5-1 provides a way to look at the needed transition.[2] It depicts two dimensions of career trajectory, business and technical. Generally, the two have been orthogonal; people moved along one or the other. The business trajectory might include some job changes such as moving from a large company into consulting or a small startup firm, but the direction was up through business and not across into technical.

The technical career trajectory until recently was just as unidirectional, the traditional career path progressing from programmer through systems analyst and project leader into technical management. The only diversification or diversion was the choice of technical specialization as in data base management, operating systems, CAD/CAM (computer-aided design/computer-aided manufacturing), or particular programming languages, or staying in the mainstream of application development.

End-user computing, office technology, and business users' often intense dissatisfaction with IS departments' lack of business understanding have turned the tidy world of IT professionals upside down. The result has been the emergence of hybrid careers that combine technical fluency with business literacy. The problem has been to find hybrid personalities to pursue these hybrid careers.

Information Systems is well ahead of most business units in this respect; however, the changes are stressful for individuals reared in purely technical environments with limited exposure to, and min-

FIGURE 5-1 **Career Trajectory Map**

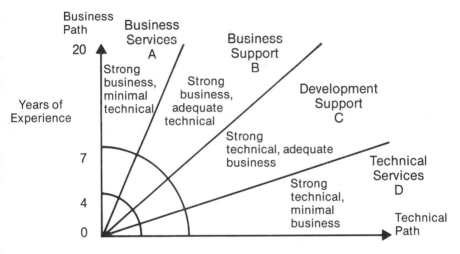

Summary of Major Roles

imal interest and training in, business. Business managers, particularly today's top executives, have largely been able to avoid IT; it was not part of their formal education, their early work experience, or their responsibility as they moved up the career ladder. Now business cannot afford technology-illiterate managers any more than it can afford business-illiterate IT professionals.

Of the four broad career trajectories plotted in Figure 5-1, the outliers represent the more traditional paths of advancement, with the inner trajectories being the trails blazed by the hybrids. We look at each of these in turn.

BUSINESS SERVICES. The essential need here has traditionally been for business skills; knowledge of or experience with IT has been either a minimal requirement or not a requirement at all. Public relations might have been an example; "might have been" because it is hard today to find any business function in which technology is irrelevant to the profile of the effective manager and professional.

BUSINESS SUPPORT. This is one of the emerging hybrid roles that will constitute the majority of management, staff, and professional positions in the coming decade. It requires strong business skills and adequate understanding of IT. In finance, this might today be an individual who works in foreign exchange but has enough technical knowledge to evaluate software and hardware and work effectively with the corporate IS department to develop dealer information systems. In IS, it might be a specialist in executive information systems who understands the information needs of senior managers and can communicate effectively with the systems development team and software experts.

DEVELOPMENT SUPPORT. Another hybrid role, this one demands primary skills in technology coupled with an adequate understanding of business. Many effective IS groups have responded to this need by establishing new "business analyst" positions and designating major business-unit users as clients who are assigned a senior IS professional to look after their needs. Information systems has become information services, and applications development has become development support.

TECHNICAL SERVICES. IS still needs first-rate, expert technical specialists. Second-rate specialists, however knowledgeable about business, cannot meet the many and growing challenges of information systems design, implementation, and operations. Some of these specialists provide essential and scarce expertise in operating systems, the extraordinarily complex software that controls the computing environment. Other technical specialties include expert systems, advanced programming tools, data base management, document management technologies, network management, telecommunications switching equipment, and local area networks.

Cost of Loss Analysis

Career options for managers who are not qualified to handle IT will be limited in the 1990s. The old assumption that the job ladder was the career path discouraged lateral movement. Through the early 1980s, applicants who had moved around a lot were viewed as "unstable" by many large firms. Experience had great value in this context. It has come as a big surprise to many displaced middle managers that this is no longer so.

It is important to know which aspects of experience are likely to increase in value and which are likely to become depreciated assets. Figure 5-2 shows "the cost of loss" for people in two very different jobs. The first, a systems programmer specializing in mainstream IBM operating systems (Technical Services), earns $50,000 per year. The second, an internal consultant responsible for helping business-unit managers select office technology and personal computer systems and educating key staff (Business Support), earns $35,000 per year.

FIGURE 5-2   **Cost of Loss Analysis**

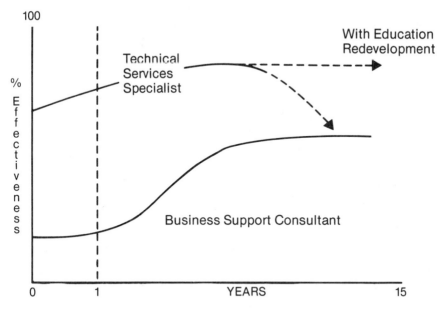

How effective is this person when hired? after 1 year?
What is it worth to keep him/her 1 extra year?
What "maintenance" is needed?

The effectiveness of each of these two people is shown on the left-hand side of the diagram in Figure 5-2. The horizontal axis shows time in years since being hired by the firm. The curve for each new employee is a rough estimate of relative effectiveness over time.

The systems programmer's initial effectiveness is high; her knowledge is generic rather than company-specific. By contrast, the internal consultant's value relies on his getting to know the organization and the organization getting to know him. In jobs that demand organizational knowledge and exposure, as much as six months might be spent learning the job and the company.

The systems programmer's effectiveness grows quickly, but drops rapidly after five years. This estimate reflects a subjective prediction that changes in hardware, software, and operating systems over the next few years will render much of her knowledge obsolete. She will be bypassed by more current, if less experienced, specialists. Education and professional development can forestall this dive.

The numbers associated with these examples are artificial, but they illustrate some of the new realities of careers in a time of continuing change. The cost of loss for an expensive technical specialist actually decreases, while losing the internal consultant at the end of year 2 means recruiting someone else and moving the person through the same learning curve. At the same time, the long-term value of the specialist may decline rapidly as a consequence of the shortening half-life of knowledge. The less expensive internal consultant, by contrast, is more costly in real terms but grows in value.

Unless firms and employees make a major and sustained commitment to continued education, technical experience is likely to become rapidly commoditized in the 1990s. The example shown, for IT experience, could just as readily represent functional-area business expertise. The middle manager in corporate banking, the accounting department supervisor, and the computer-illiterate financial manager are equally susceptible to premature erosion of the value of their experience. For their part, firms are likely to underestimate the fairly extensive nurturing needed to generate the combination of technical aptitude and knowledge and organizational learning that characterizes the hybrid.

Cultivating Hybrids

Through the early 1980s, the hybrid career trajectories of business support and development support were largely empty; in most organizations, people were moving up through either business services or technical services, and individuals with the requisite skills were hard to find. They still are, and they also face major problems getting promoted.

Consider, for instance, the increasingly typical example of an entry-level person in finance who becomes interested in personal computers. He learns how to go beyond standard spreadsheet applications, perhaps building new credit-risk analysis models, helping to install a local area network, and designing a departmental information systems facility. After a year or so, he has become indispensable but, no longer viewed as in the mainstream of finance, not promotable. Bypassed by peers who have gained experience on financial projects, he applies to the Information Systems department but lacks sufficient technical background for the specialized positions there. Eventually hired by a consulting company, he brings a combination of business smarts, technical creativity, and personality that prove to be just what clients trying to use computers in their business need. The client firm contrasts his skills repertoire with those of its own finance people, who lack technical skills, and its IS people, who tend to be unresponsive to business needs.

A similar example of a hybrid caught between two cultures is the IS entry-level programmer who has a reputation in finance as the person to talk to for help with desktop publishing, electronic mail, and word processing. She gets several fast promotions, but after two years is stuck in her job, which amounts to little more than being a support desk for personal computer information. Her supervisor criticizes her lack of project management experience. She, too, eventually leaves, and within two years hires three of her old IS peers and her boss to join her fast-growing office services department in a publishing firm, where she now earns three times her old salary.

Few companies consciously try both to build a cadre of hybrids and to provide them with real career opportunities. Service-oriented IS departments that do recognize these unusual people as a key resource in bridging the culture gap between the world of business and IT are unsure how to recruit, grow, retain, and promote them. Almost no firms require their "fast trackers" to cross the gap on their way to the top. Consequently, today's hybrids find themselves in a twilight zone of career ambiguity. Management could change that simply and fairly quickly by making a brief trip—six months to two years—across the culture gap a requirement for fast trackers on their way to the top.

Lateral Development

It is easy to demand that business and IS executives be fluent in both IT and business, but several decades of separate cultures, career ladders, education, and experience mitigate against this. In some firms, business executives are becoming familiar with IT through a form of organizational osmosis. For example, a bank's head of retail marketing responsible for point of sale, credit card products, and customer accounts is, or should be, in close contact with the head of computer operations and key systems development projects. Similarly, airline marketing and operations managers learn about IT as a result of their responsibility for aspects of the business that depend on computerized reservation systems, as do financial executives as users and sponsors of corporate cash management and treasury systems.

Such people are at least literate about IT and comfortable with becoming involved in planning and decision making. They often move into IS for a few years, some becoming heads of IS. But very few organizations make such lateral movement a routine part of management development, and IS has not been an attractive initial assignment for anyone targeting the top. In most firms the number of senior business managers who started out in IS can be counted on the fingers of one hand! Is it surprising, then, that the annual lists of which functional areas Harvard and Stanford MBAs are entering rarely show even 1 percent going into IS?

To develop the hybrids who are already so badly needed, firms must make crossing the "culture divide" a requirement for advancement. If 10 percent of each entry-level group of managers were to make the shift for a period of six months to two years, the problem would solve itself within five years.

The process would work something like this. The move would be made about two years after entry to allow for the development of experience and background in the individual's area of focus. The assignment would be limited to a maximum of two years; any longer and a person would find returning to his or her original business or technical function difficult. Return would be required, the desired value of lateral development being to develop the skills needed to bridge the culture chasm and facilitate collaboration across historically separate parts of the organization.

In the absence of lateral transfer as part of a firm's formal policy, the only practical option for those who would be hybrids may be to find other routes to cross-cultural skills development. If they want to stay with the firm, they might try to get close to the operations and back-office functions. IT can be expected to significantly blur the distinction between the back office and front end of the business in the 1990s, and those who understand core operations will know how the firm's delivery base and coordination systems really work. Working on a large systems development project—one that involves a team of at least ten people over a period of at least two years—is a real learning opportunity for anyone who wants to move along a business services trajectory. Familiarity with small-scale technology and its applications affords surprisingly little real insight into the complexities of IT design, development, and use. Small-scale systems are to large-scale systems what bicycles are to air transportation; bicycles are wonderful, efficient devices, but they do not pose the same challenges as air transportation in terms of infrastructure, complexity of components (airports, maintenance, route scheduling, safety), and specialized skills.

If these options are not available through staying with the firm, joining a management consulting company for a few years might be the most effective way to develop hybrid skills. Such companies often cultivate an environment that encourages rather than limits cross-functional teams, and their clients are usually looking for breadth of background.

A "Skills Scoreboard"

The development of the hybrid manager can be characterized in terms of the balance needed between two types of skill and experience. The required business skills will include business and functional skills and also personal and organizational skills. The former skill area encompasses knowledge of a relevant business area in a specific industry, functional specialty, or specific firm's overall range of activities. This area of skill is heavily dependent on experience. The latter skill area includes the ability to communicate, to facilitate effective collaborative relationships, and to build credibility outside one's area of specialization. The requisite technical skills rely on breadth of experience with applications, types of systems, and other technical tools and currency of specialization. As information technology covers a wider and wider range of ever-changing technical building blocks and tools, breadth of experience often needs to be complemented with up-to-date knowledge of data base management systems, local area networks, image technology, advanced software development methodologies, and so forth.

To illustrate how specific skills requirements can be identified, imagine building a point-of-sale capability in a bank. Just about every bank retailer, automotive firm, petrochemical firm, and distributor will need to develop comprehensive POS products and services within the next five years. This example suggests the human capital resources that must be deployed to do so.

Responsibility for development of the bank's POS capability will rest with four individuals. A network specialist will supervise the design and operation of a technically sophisticated telecommunications, computing, and information management resource that must operate at several hundred transactions per second, maintain levels of reliability well above 99.5 percent, and provide consistently fast response times for the merchants who use the service and for automated network management and security. The bank will also hire a product development manager, a new position, whose primary responsibility will be to work closely with retail customers to de-

FIGURE 5-3 **Scoring the Skills Needed for the Four Positions in the POS Development Project**

	Network Specialist	Manager of Applications Development	Product Development	Financial POS Specialist
Business skills				
Business/functional	1	3	5	5
Personal/organizational	1	3	5	3
Total	2	6	10	8
Technical				
Breadth of experience	4	5	3	1
Currency of specialization	5	3	3	3
Total	9	8	6	4
Role quadrant	**Technical Services**	**Development Support**	**Business Support**	**Business Services**

velop a range of POS services, including electronic funds transfer at point of sale, direct on-line debiting of credit card and bank accounts, and merchant information systems tracking consumer purchasing patterns. An application development manager will guide the modification of existing software systems and the design of new ones needed to provide the POS products and services. Finally, a financial specialist will manage the bank's pricing strategies for POS and be responsible for profit planning and financial forecasting and reporting. A simple scoring system is used to assess the mix of skills required for each position (see Figure 5-3).[3]

The network specialist must have extensive experience in large-scale, on-line transaction processing systems and be familiar with advanced telecommunications switching systems and POS technology products and their vendors. The correspondingly high scores for technical skills, coupled with low ones for business and organizational skills, put this job in the technical services quadrant in Figure 5-1. Sound business experience—knowledge of banking and outstanding communication skills—would not compensate for deficient technical skills of a candidate for this job, but neither would the person have to be a hybrid.

The product development manager must possess a sound knowledge of banking and retailing and will be required to work closely with retailers and almost as closely with the IS application teams that will turn the product designs into software. He or she must be able to effectively communicate customer needs to IS and relate the technical issues to business issues. This position scores extremely high in business skills and moderately high in technical skills, putting it into the business support quadrant in Figure 5-1. The desired candidate is thus a hybrid. The bank may find it difficult to locate such a person. Someone from marketing who spent time in operations or represented the user community in systems development projects, if there were such a person, would be a strong contender.

The manager of applications development will need to possess extensive knowledge of POS practices and operations in a business context as well as POS technology. Merchants' needs and operations must be accommodated—the bank cannot impose its own procedures—and all who interact with the system must find it easy to use. Consequently, the manager must be able to communicate well both with clients and with the bank's business teams. Twenty years ago, a strong candidate for this position would have ranked extremely high in technical skills and very low in business skills. Today, the scores are almost on a par, slightly higher on the technical side, putting the job into the development support quadrant in Figure 5-1. Again, a hybrid is vital. And again, finding someone with broad technical experience who is reasonably technically current, moderately knowledgeable about the business, and who possesses good communication skills may prove difficult. Breadth of experience will be vital, since POS systems development addresses a broad range of computing, telecommunications, and data management issues. Unfamiliarity with on-line transaction processing systems will be a disadvantage. A person who satisfies most of these qualifications is unlikely to have been able to keep current in all areas of technical specialization but, being broadly aware of major trends, can rely on technical specialists for advice.

The financial POS specialist must understand aspects of the technology that significantly affect business costs. For example, design of the telecommunications network and choice of vendor for telecommunications transmission may depend on projected transaction volumes. Using public networks may be initially attractive when volumes are low but will significantly degrade profits if business takes off. The need to be knowledgeable about such trade-offs warrants a low to moderate score in technical skills, while business skills rate high, putting this position into the business services quadrant in Figure 5-1. The strong candidate here need not be a hybrid.

Zero-Based Work

Experienced practitioners and academics agree that many of the disappointments and problems of IT come from automating the status quo rather than rethinking, streamlining, and eliminating work. IT is most effective when it redeploys human capital—when it cuts out unnecessary bureaucracy, leverages skills, and transforms the quality of work in an organization.

If jobs stay the same or are tinkered with at the margin, innovation will be damaged. If jobs change but people either will not or cannot, stress, alienation, and incompetence will result. If jobs and people change, but not managers, expect a loss of trust and respect. How frequently these things occur and how damaging they can be are amply illustrated in Shoshana Zuboff's *In the Age of the Smart Machine*.[4] In example after example, Zuboff documents the failure of management processes to adapt to the opportunity to build a "learning organization" and to leverage people and technology jointly, processes she terms "informating" to distinguish them from simple automation.

A striking example of the impact of rethinking rather than automating work, an example that suggests that if a job is not worth doing, it should not be done, is provided by Ford's system for paying suppliers. Ford cut white-collar staff for eight consecutive years between 1979 and 1988, largely by substituting IT for people, and in the process transformed itself from an exemplar of just about everything that was wrong with U.S. firms into a model of competitiveness.

Ford's standard supplier payment procedure involved a clerk, upon receipt of an invoice, phoning the appropriate plant to confirm the arrival of the parts and verify the price. Payment was then authorized or a lengthy checking process begun. When it initiated a project to computerize the process in 1986, Ford's goal was to cut 100 of the 500 jobs in the supplier payment function.

"Then someone said," according to Ford's head of Information Services, " 'Why pay bills in this cumbersome way at all?' " Today, the document attached to the arriving parts is fed directly into a workstation at the receiving dock. A central computer checks the figures and immediately authorizes a check-writing machine to make the payment. Suppliers do not send invoices at all now. Instead of cutting out 100 of the 500 jobs, Ford has eliminated close to 400.

Ford's achievement may sound simple, but it wasn't. The company had to bring together people from manufacturing, finance, purchasing, and data processing to change basic work patterns. For example, purchasing agents who used to negotiate prices with suppliers now are also responsible for ensuring the prices in the computer's memory are up to date. "The hardest job is crossing organizational lines."[5]

Today's jobs are irrelevant to tomorrow's work. The people in them are not.

Education: Maintenance for Human Capital

Training follows technical change; education should lead it. Education is preparation for moving into a new context. Generally, firms do too much training too late and don't provide enough education. For example, the introduction of personal computers is often accompanied by hands-on training in how to use a software package or electronic mail system. The attendee learns mechanics and procedures. There is rarely a broader education program that looks at where the personal computer may fit into existing operations, how it is intended or expected to change work, and where the applications fit into the broader business and organizational context.

The need for IT education is immense. Hardly any major firm has not at least begun management "awareness" programs. Chief executives are joining their top five to ten managers for a precious and hard-to-schedule session of two to three days to be briefed on IT. Many firms put their top two hundred managers through a required education program. Others are sending their IT staff to courses that help develop their business and organizational skills. Special workshops aimed at broadening Information Services departments' perspectives on their role within the organization are becoming common. Education is the fastest-growing component of the IS budget in many firms, albeit having started from a small base.

The business need for education is stated clearly and typically in a planning document drafted by one of the world's top banks.

> Less than ten years ago, very few managers in any large organization had experience with information technology or much need to know about it. The Bank was no different.
>
> The challenge facing the Bank and its managers is to take full advantage of the information technology already installed in the Bank, to exploit the potential advantages [it offers].
>
> To meet that challenge, Bank managers and staff must be educated systematically and on a continuous basis. . . . [They] have to become more familiar with both the potential and the pitfalls of information technology and more committed to its effective management.

Imagine reading such a statement ten years ago. Most senior executives would have wondered what the fuss was all about. Now, they accept the fact that IT education is essential. That said, they often see it as needed by others, not themselves. In this bank, the people at the top still largely avoid getting close to IT issues. They are at a stage of management awareness but not management action. Education is a threat for them as well as an opportunity for the organization.

As a force for change, education is not an event, an occasional course away from home taught by a well-known professor or consultant. Firms need to think of education as a capital investment, to be planned carefully, evaluated in terms of business return, and developed systematically. The central element in any education strategy is not the content of the program, but its behavioral objectives—not what people learn from the course, but what they do as a result.

Objectives of IT Education

The priority of most management education programs during the 1980s was to increase awareness and thus open a dialogue between business and technical people. As we have seen, awareness is the essential starting point for, but no guarantor of, action; it can as readily lead to delegation.

For many managers today, the problem is not lack of awareness or unwillingness to take action but a lack of the vocabulary and skills needed to participate in policy and planning. Many senior managers do not know the issues on which to focus. They don't know what makes a sound IT plan. They don't know what "ISDN" and "EDI" are. They do need explanations that will stimulate them to take action.

Education programs that bring together in a supportive context people from both sides of the IT business culture and knowledge chasm can provide a forum for airing concerns about costs, politics, timetables, jobs, and the like—issues that too easily go unexpressed and unresolved in the formal planning process.

The gulf between business and IS is hinted at by the abstract and somewhat dismissive term "users," widely adopted by the IT profession. Education can contribute to the heightened collaboration that most commentators on implementation see as vital: technical designers and their business colleagues and clients working jointly together in a process that combines technical and organizational change.

When IT is used to significantly alter the status quo, as it was at Ford, entirely new dialogues and education are needed. Many management education programs have no clear behavioral objectives. Such programs often leave their audiences bewildered. The consequence is that in the second decade of the personal computer, information technology remains an intimidating subject for most business managers. It is all too easy to confirm their expectations that IT is over their heads, technical, boring, and irrelevant. It is extraordinarily hard to avoid jargon in IT education, and harder still to know when to explain the technology and when to avoid discussing it.

Who Needs IT and a Strategy for Delivering It

Every business manager over 30 needs some form of IT education. That is the opinion of most over-30 IS managers. All over-30 Information Services managers need business education. That, alas, is often not their opinion but that of their business "users." Both sides of the IT/business culture chasm accept that education is a key to bridging it. Each party has a clearer idea of what the other should learn about than what its own education priorities should be. Information Services staff wish that senior managers would understand the realities of systems development, the long lead times involved, and the importance of the corporate architecture. Business executives wish that the IS staff would learn to communicate, think in terms of service instead of technical elegance, and listen, not lecture. The point of agreement is that for a firm to benefit rather than suffer problems and frustrations on account of IT, education is essential.

Figure 5-4 outlines the basis for an education strategy that addresses the concerns of all parties in the organization concerned with or affected by IT. The strategy is that management education programs need to be action-oriented rather than topic-based. The organizing message to managers should be:

- Know where you will be affected by the growing interdependence of business and information technology;
- Know the major decisions and responsibilities that will fall to you; and
- In terms of the technical background that you need, understand that the question is not how much you need to know, but the

FIGURE 5-4 **A Strategy for IT Education**

Identify priority communities.

The different communities must be identified at the start, together with the sequence of priority.

Business managers:

> The top management team (typically, 5–12 executives)
> Senior management (50–100)
> Middle managers
> Supervisors
> Staff
> Entry-level employees

Information Service staff:

> Managers
> Technical Services roles
> Development Support
> Business Support
> Business Services

> Should the starting point be education for policy action at the top? Education for implementation in the middle? Business and organizational education for IS professionals?

For each community:

> 1. Define operational goals for action, not topics for edification. What people should go away thinking, feeling, and doing is as or more important than what they should go away understanding.
> 2. Identify "so what" issues.
> People need to know why they are being asked to learn and why and where IT will affect their roles, work, decisions, and careers.
> 3. Determine the degree of customization needed.
> Is off-the-shelf material adequate to meet operational goals? Where must IT be thoroughly and concretely related to the details of the organization?

Commit resources for at least two years.

> This recognizes that education is not an occasional event but a continuous process, and not a gentle time out from work but a serious and often disconcerting part of a program for organizational change.

Design and deliver.

Follow up and sustain.

minimum you must know to be able to play an active role in your firm's deployment of IT.

The first consideration in an education strategy must be the recognition that each management community has entirely different education interests and needs that match the different impacts of IT on their work and career directions, decisions, and responsibilities. The principal communities are the top management team, senior managers, middle management, supervisors, entry-level employees, and the Information Services organization.

Education must be customized to the environment and context of each of these communities. Off-the-shelf, generalized courses, though easier to develop and cheaper to teach, may leave attendees better informed but none the wiser. Managers in most companies have very little idea what their firm's IT base and strategy are. An education program that treats these subjects in grand abstractions can frustrate managers who want to understand their own technology and strategy and how it all fits together.

Any serious education effort requires at least two years' commitment of funds and time and should not be confined to just one level of the organization. Occasional courses targeted at only a few communities do not build a critical mass of shared understanding. For mid-level managers to be able to take a more active role in selecting IT opportunities within their own business units, top management needs to have reviewed policies in areas such as business justification, need for a corporate architecture, and funding. Otherwise, old policies and processes will block new initiatives built on new learning.

Education should be viewed as a forum for building dialogues for planning and implementation. In many companies, business managers rarely meet as a group to review the basics of IT as a business resource for the firm. If they meet with IT staff, it is usually for budget meetings, project reviews, and other activities strongly focused on specific tasks and agendas. An education program can foster the kind of open and open-ended discussion needed to bridge the culture gap between business management and IT professionals.

The economics of IT scare almost all managers. IT is the only major business function that continues to increase at a significantly higher rate than revenue and profit growth, and whose benefits too often are unproven. Education about IT must include enough about its economics to enable managers to begin to quantify costs and benefits.

The technology itself must be appropriately addressed, even when the main focus is on competitive issues and business management. It is extremely hard to make the technology relevant to business managers. It is even harder to find a level of detail that avoids the two extremes of being misleading and oversimplistic on the one hand and too technical on the other.

The current fashion in most management education is to avoid the technology entirely, which gives an incomplete perspective. So many critical decisions concerning business uses of IT relate to standards, integration, the telecommunications infrastructure, and the technical features of the corporate IT platform that managers must be familiar with their importance and thus with what they are. If IT planning was overtechnical and ignored business priorities through the 1970s and early 1980s, the pendulum swung too far back in the late 1980s. Technology, however complex and varied, is the enabler. It is time to put the "T" back in IT and competitive advantage.

It will not be done with off-the-shelf programs from universities or consulting firms that rely on standard business-school cases of competitive advantage. This approach was fairly effective at building management awareness. The need is now to turn awareness into action. We have seen that delegation is the usual consequence of awareness without a business vision, or even *with* a vision if there is no compelling message about the need for a corporate IT platform. It is for this reason that management education must address the technology.

The varying education needs of the different management communities are summarized in Figure 5-5. The CEO and top management team, concerned with long-term competitive and economic issues, for example, need education that addresses funding, major new infrastructure investments, and overall policies for managing the platform and applications. The competitive context, the relationship between business-unit and IT planning processes, and historical problems with implementing major new systems cost-effectively and on time are concerns of senior managers at the next level down. For these managers, many of whom view the corporate IS unit as an unresponsive blockage, how much to learn about IT can become a personal quandary.

Middle management's education needs are far more specific. These managers do not directly affect corporate policy nor, gener-

FIGURE 5-5 Education Priorities for Organizational Communities

	Communities					
	Top Managers	Middle Managers	Technical Managers	Development Support	Business Support	Users
Managing IT: Policy issues, key decisions, resource needs, competitive priorities	X					
Managing the economics of information capital: Cost dynamics, trends, measuring business value	X		X			
Introducing IT: Basics, demonstrations, presentation of the firm's IT direction and its implications		X				X
Building systems: Vocabulary, methods for collaborative design, user/technical staff roles, project management				X	X	
Implementing systems: Managing organizational change, forum for open expression of concerns				X	X	X
The new IT environment: Emerging technologies, applications and high-payoff opportunities			X	X		
X for non X-ers: e.g., marketing for non-marketers; knowledge of business, exposure to what user community does and cares about; building consulting skills				X	X	

ally, are they interested in grand, competitive strategy. They tend to see IT costs as a growing burden and want to know how to realize fairly immediate, short-term, bottom-line benefits. Middle managers are much more likely than their seniors to be interested in demonstrations of new tools and systems that will provide a concrete sense of the technology coming into their world.

Information Services professionals expect to be kept up to date through ongoing technical training. They also expect to provide technical training to their users. But few expect to be required to attend education programs that will extend their business understanding and organizational skills. This is the education that will enable them to shift onto business-support and development-support career trajectories. Even those who choose to remain in a technical services role will need frequent training. A rough rule of thumb is that technical specialists in any fast-moving field who spend less than 10 percent of their time on professional development are likely to fall behind.

The specifics of any management education program customized to the context of a given community must be defined. Yet more important than the details are the principles of education for action, sustained commitment, and a forum for dialogue. Management IT education is today a crucial component of any effort to mobilize to accommodate accelerating change.

The Role of Human Resource Managers

One of the most effective alliances in any organization that acknowledges IT as part of its main business thrust is that between enlightened senior Information Services and human resource managers. Together, they can create an education plan that drives rather than follows change.

In the absence of such an alliance, HR managers, to whom "training" budgets belong, regard decisions on IT education as their prerogative and see IS intruding on their turf when it brings in outside consultants or proposes a series of programs for top managers. In practice, IS must play a leading role in senior management education, not to teach managers "about" computers but to provide a grounding for rethinking IT planning, policy, the business/IS dialogue, and competitive and organizational needs. In many ways, such education is a disguised form of organizational development. In an effective alliance, HR will play a major administrative and logistical role and IS will be heavily involved in design and development and participate in classroom sessions.

IS should not move on its own. Education is not for amateurs or for consultants with limited teaching experience. HR has the professional experience to ensure that programs are clear in their behavioral objectives, structures, and teaching points, and incorporate follow-up and appropriate evaluation criteria. To this, IS can bring the background information and planning agenda needed for effective education.

Management action here is simple: make sure the alliance is built. It is not easy to redefine roles, build a cadre of hybrids, formalize lateral development as a necessary requirement for fast trackers, and change incentive and reward structures. The human resource function must lead these activities.

IT is changing the basics of jobs and work, and HR policy must adapt. This presupposes an understanding of the new skills and jobs by HR managers. Evaluating salaries and promotions for IT jobs in relation either to comparable grade levels elsewhere in the organization or to published averages for IT job categories can be like asking what the average salary is for a lawyer. The answer might be $58,000. But that tells us very little about either lawyers or which lawyer we should hire to deal with the unpleasant problem of an antitrust suit, nationalization of our assets in Outer Farawaya, or the unfortunate arrest of the head of marketing at the Las Vegas Cordwangler Manufacturers Convention. The new jobs and careers IT is creating—or rather the jobs that the effective use of IT increasingly depends on—do not fit comfortably within established grades, salaries, recruiting practices, and job ladders.

A 1987 study of human resource issues in office technology concluded that few HR managers see a role for themselves in setting technology policy.[6] The majority can thus be expected to be reactive, becoming involved only after implementation, mainly to deal with consequences of the new system such as increased turnover, job dissatisfaction, stress, and a mismatch between available and needed skills and resources.

HR could play a strong, supportive role in terms of job redesign and reclassification, education and training, and policies for handling attrition. Yet even this would be deficient by failing to look far ahead and understand the issues of organization redesign and the new career trajectories. Ideally, HR should strive to identify "what the 'ideal' work structure should be, from both a social and technical point of view," adopt an "opportunity orientation . . . which explores the possibility that system design and implementation activities may be used to move forward toward preconceived organizational ideals and preferences," and develop "an organizational impact statement."[7] This more active approach relies on an understanding by HR managers of the wider business and competitive context of IT and on IT managers' appreciation for the human context of the business.

Skilled and active HR specialists add to the business-IT dialogue an understanding of many other aspects of human capital redeployment that are not addressed in this chapter only because there are so many. These include strategies for effective implementation, team-building, encouraging entrepreneurship and intrapreneurship, compensation and incentives, employee relations, productivity measurement, and supervisory and management processes.

In addition to dealing with these fairly action-focused concerns, HR can contribute to the debate that will surely persist well into the next century over whether IT will improve or reduce quality of working life, employment levels, autonomy and sense of worth in jobs, relationships between management and workers, trust, and job satisfaction.

Notes

1. For background, see Shoshana Zuboff, *In the Age of the Smart Machine* (New York: Basic Books, 1988); R.E. Walton, *Up and Running* (Boston: Harvard Business School Press, 1989); T. Forester, ed., *Computers in the Human Context* (Cambridge, MA: MIT Press, 1989); and R. Howard, *Brave New Workplace* (New York: Penguin, 1985). These books range from the cautionary (Walton) to the downright pessimistic (Howard). They are all well researched and compelling in their evidence; they are a valuable antidote to the massive literature on the wonders of IT that entirely ignore human issues.

2. This model of career trajectories is presented in more detail in P.G.W. Keen, "Roles and Skill Base for the IS Organization," in J.J. Elam et al., *Transforming the IS Organization* (Washington, DC: ICIT Press, 1988).

3. Ibid. (for more detailed examples and applications of the model by firms).

4. Zuboff, *In the Age of the Smart Machine*.

5. Quoted in C.L. Harris et al., "Office Automation: Making It Pay Off," *BusinessWeek*, October 12, 1987.

6. R.J. Long, "Human Issues in New Office Technology," in Forester, *Computers in the Human Context*, pp. 327–334.

7. Ibid., pp. 330–331.

Chapter 6

Managing the Economics of Information Capital

When a company budgets $1 million to develop a new software system it is, in fact, committing to spend more than $4 million over the next five years. Each dollar spent on systems development generates, on average, 20 cents for operations and 40 cents for maintenance.[1] Thus, the $1 million expenditure *automatically* generates a follow-on cost of $600,000 a year to support the initial investment. Development is in many ways the loss leader for maintenance.

Software development is the main discretionary expenditure in IT planning. It is also obviously the key element in using IT for business innovation. To cut back on software investment is to risk cutting back on business development. But software is extremely expensive when measured in terms of full lifecycle costs. To ensure that the true costs of software are factored into business justification and that the capital provides real returns requires major shifts in the management process for IT.

For each of the past three decades, IT budgets—for hardware, software, and telecommunications—have grown at about 15 percent per year, which is far greater than the rate of business growth. It will be impossible to sustain this rate in the 1990s. IT expenditures of a billion dollars a year by leading firms have brought this area to the forefront of capital planning. In financial services, for example, IT operations costs alone are the third or fourth largest expenditure, behind employee costs, real estate, and interest. A growing number of senior executives are worried that IT costs may be out of control.

IT now amounts to about half the incremental investment for large firms. It has become a contender for scarce business capital and not just for tomorrow's expense budget. There is little evidence that the investment is producing adequate benefits. What is more, there are no reliable methods for measuring the business value of IT.

Senior executives are caught in a worrisome double bind: ever greater commitments to IT investment are being driven by competitive necessity and discouraged by escalating costs and uncertain benefits. Put another way: economically, companies cannot afford to increase capital spending on IT; competitively, they cannot afford not to do so. The economics of information capital is firmly on the top management agenda, and corporate managers are clamoring for help.

There are three issues in managing the economics of information capital: managing costs, managing benefits, and managing risk exposure. That these have generally been handled poorly is attributable almost entirely to the historical emphasis on treating IT as overhead, which is managed through budgets, cost allocations, and cost justification. It is a very naive way of dealing with a complex economic good.

In managing the benefits, making the business case, and assessing payoff in relation to anchor measures that relate to operational business indicators, management must accept that there exists no set of accounting ratios or simple formulas that show the business value of IT. In this respect, IT is like R&D; it must be justified as a longer-term investment in the future.

Finally, managing the risks associated with IT expenditures involves market concept risk, technology risk, implementation risk, economic risk, and organizational risk.

Managing IT Costs

Managing costs should begin with the creation of an IT asset balance sheet and commitment of time, attention, and management resources appropriate to the amount of the capital asset (most managers will be extremely surprised by its size). Management must then count all the IT costs, many of which will be hidden. Development compounds future operations and maintenance costs, and organizational, support, and infrastructure costs are frequently overlooked or obscured by an accounting system that expenses IT as overhead.

Creating the IT Balance Sheet

Creating an IT asset balance sheet that capitalizes all IT equipment, software, and data resources currently in use is one way to jolt senior management. The balance sheet is not a legal financial statement—accountants point out that software cannot be capitalized for tax purposes—but instead a management report. Management should know what the annual IT expense is, but it very likely does not know how much capital is tied up in IT resources. It is probably far more than management realizes. And it also often comes as a surprise that there is no way to tell from the accounting system; because software development is expensed, it is often impossible to accurately calculate the value of software in use.

A balance sheet for a large bank that spends about $200 million a year on IT is shown in Figure 6-1. The bank has long been aware of the most obvious capital item, hardware and equipment, but has overlooked the capital tied up in personal computers, departmental telecommunications, and small unit-cost items that together added up to several million dollars.

Top management had no idea that the bank had spent close to $500 million to create the software currently in use, nor that it was an information factory sitting on data that had cost well over $1 billion to create (this figure includes relevant fractions of the salaries of staff in the bank's operations and back-office units). IT assets added up to over $2 billion.

The exercise drew top management's attention to several points that should have been obvious, but that are almost always overlooked when IT is expensed. The main point was that IT assets were undermanaged. In this firm, corporate finance has five times the number of senior managers as Information Services, which makes do with a senior vice president and three subordinate VPs. The amount of time spent on IT at top management meetings was measured in minutes per year. Partly as a result of understanding the level of capital IT represents, the top management now commits days per year to IT.

This is not a call to treat IT as "special," but merely to provide a management complement appropriate to the level and business importance of the asset base. Many executives view the trend of creating chief information officers (CIOs) as a ploy by IS to boost salaries. If IT were treated like any other business unit that manages $2 billion of assets, top management would surely elevate the level,

FIGURE 6-1 **An IT Asset Balance Sheet**

	Assets ($ millions)	
Hardware:		
Centrally managed computers	120	This is the most obvious component of the IT base and the one that accountants track; it is just 5% of this bank's real IT assets.
Distributed computers	84	Mainly personal computers. In many organizations, PCs, workstations, and departmental systems now account for more expenditures than do central expenditures.
Network equipment	105	Telecommunications facilities, often distributed across many different operating budgets.
Distributed telecommunications	59	Local area networks and departmental equipment.
Total hardware	368	
Facilities (data center and operations)	192	
Software:		
Application development	420	Software development expenditures are expensed. The bank had no idea how much it had spent to build the software in use, nor did the accounting system make it easy to find out. This and the figure for "other software" are really educated guesses. It also ignores replacement costs, which the firm's IT planners estimate as at least $1.2 billion, three times the original development cost.
Other, including personal computer software	68	
Total software	488	
Data resources	1,200	This is the estimated capital cost of the salaries, processing, and storage incurred in creating the on-line data resources that are the basis for the bank's products and services, which is an infinitely reusable asset. Data resources do not wear out as they are used.
Total assets	2,248	

number, and quality of senior managers and planners assigned to it.

The $490 million in software and $1.2 billion in data currently in use in the bank are already paid for. Can it be more effectively used and reused? Can new products be derived from it? Can the software be repackaged?

Of course they can and should. The organization now sees many opportunities to reuse its data and to realize direct cost savings by, for example, merging decentralized facilities into the corporate IT utility. Additionally, powerful new mechanisms have been established to ensure that corporate staff, senior management, and line managers spend more time and attention not on IT as such, but on the policies and priorities most relevant to the biggest single element of capital in its fixed-asset base.

Before the firm built its IT asset balance sheet, no one had noticed any of this. Even IS was surprised by the size of the software and data capital asset.

Counting All the Costs

When IS was a central corporate staff function, its costs, if they could not be controlled, at least were relatively easy to identify. Today, with almost half of firms' IT expenditures going for personal computers, office technology, and end-user computing, even identification has become much harder. Many companies do not know what they are spending on IT. End-user computing and telecommunications expenditures are scattered across business units' budgets and many organizational costs are not even tracked. *Datamation* estimated in 1987 that 40 percent of the costs of IT were not part of the Information Services function's budget.

How can firms make rational decisions about IT when they do not know its true costs? The starting point for managing the economics of information capital is to understand that IT costs are of two types: supplier costs (i.e., the costs incurred by the Information Services function) and user costs (the growing portion of firms' total IT expenditures that is being shifted out of central corporate IS units into the business).

SUPPLIER COSTS. Corporate IS groups are a form of internal supplier, analogous to a utility, that operate corporate data centers and telecommunications facilities and provide services such as systems development and information management. Their costs are usually charged back to the business users they serve through an allocation mechanism equivalent to transfer pricing in manufacturing. Computer hardware and operating systems and telecommunications equipment comprise the power plant for information services. Frequent changes in the components of this power plant, driven by rapid technological change, require continual retraining of the people who operate and maintain it.

An increasing proportion of the operations function is committed to ensuring the reliability, security, and availability of key business services that depend on the IT base. When an airline's reservation system is down, so too is its business. When a bank's cash management system is not secure, it opens its vaults to thieves, eavesdroppers, and passersby. When a dealer order-entry system is slowed by overloaded computing or telecommunications facilities, service, cash flow, and customer image are degraded. Operations skills and resources are essential to the on-line business enterprise and a vital element in the technical infrastructure. They are also not cheap.

The business services delivered via IT hardware are realized in applications systems. Development, and subsequent operation and maintenance, of these systems is also people-dependent. Whereas systems development costs are discretionary, subject to increase or decrease by management at short notice, operations and maintenance costs are not. Inasmuch as maintenance can amount to one to three times development cost, today's development budget sets the IT budget for the next several years. The head of Information Services in a major bank calculates that at any given time his group has close to one hundred ongoing maintenance/enhancement/upgrade projects for the bank's checking account services, which drastically limits his unit's ability to deliver new systems that business groups urgently need. "They see us as unresponsive and not meeting their needs," he explains, "but it's old systems that dominate our budget and schedules, and I have no way of changing that for at least the next five years."

For every dollar of initial development expenditure for a large system, operations costs will average 20 cents per year and maintenance 40 cents per year. More staff will generally be working on maintaining and enhancing old systems than on building new ones. At any given time, only about 10 percent of a corporate Information Services unit's staff is developing new systems; maintenance occupies 50 percent or more of its scarce human resources.[2]

The cost of support is growing rapidly as IS groups shift from building systems to supporting business units. The support facilities for a $5,000 personal computer, for example, typically amount to at least $8,000 per year according to the Gartner Group, a leading research organization that tracks IT costs and trends.[3] Much of that will be paid for by the business unit, but it is a new burden on IS units, and one that they must take on if they are to be a responsive service unit. New expenses such as these are added to the specialized technical staff needed to manage the IS group's telecommunications, data management, and computing infrastructures. All these resources provide education, training, and internal consulting throughout the organization.

On the supplier side, only new development is truly a variable cost. Data centers, telecommunications networks, and operations and maintenance include variable components, but it is hard to cut overall costs except at the margin.

USER COSTS. User costs vary widely, depending on company policy regarding IS cost allocation and recovery. Some of these costs, being hidden, are unbudgeted.

Allocations and charge-back of central IS unit-supplier costs obviously become user costs. To these are added direct acquisition and usage costs. Personal computers are a frequent element of the former. Direct usage costs, which largely depend on transaction volumes, can grow rapidly and be hard to control. For example, a single personal computer accessing an outside information service may not need special justification or budget. But with IT, supply often creates demand. The success of the initial system may stimulate rapid expansion. Two hundred personal computers accessing the outside information service will shift the cost dramatically from overhead to capital, especially if it is decided to link the computers in order to share information, messages, and software.

There is an obvious and immense difference between initial small-scale, ad hoc computer use and wider departmental capability. To manage such investment case by case (piecemeal) is to likely overlook how a series of $5,000 personal computer purchases can become a million-dollar capital plan just for hardware acquisition. And the hidden costs of support and telecommunications can dwarf this expenditure.

Central IS units have long struggled to manage compounded costs driven by software development. Business units now face similar stresses.

In the days of central mainframes and IS monopolies, business units had no need to develop technical and operations staff. But with the advent of personal computers, distributed development, office technology, departmental operations, and end-user computing, these units began to incur many of the costs of distributed information management previously borne by the central Information Services group.

The apparently low costs of distributed hardware, such as personal computers, and of software packages, such as spreadsheets and word processing applications, has belied the often substantial costs of support. Organizational costs—including management time, the learning curve for staff, education, and other costs of making the transition from old work systems to new—are rarely budgeted and can make official development costs look like pennies.

IS/Business-Unit Perspectives on IT Costs

Business units often do not see the scale or value of central IS infrastructures. They do see the money they are charged for these resources, whether directly for usage or through an allocation for their part of the shared base. Similarly, many business-unit managers view as an expense, rather than as the essential service asset that they are, the legions of systems programmers, application developers, analysts, operators, and project managers who develop the software and information management asset that is so critical to competitive positioning and a basic element of efficient business operations.

For their part, IS managers have no ready formula for allocating infrastructure costs fairly. Consider this representative sample of the questions they face. Should the first users of a telecommunications service carry the full burden, even though the marginal cost is low? Who should pay for the development of a customer master data base, an infinitely reusable resource that will eventually be of value to many business units? Where outside services are cheaper, should business units be allowed to use them and thereby reduce the customer base for the shared corporate resource? With supplier costs being increasingly fixed investments that may not pay off for many years and development only a fraction of the full lifecycle cost, how can the true costs of a service be determined?

Supplier costs, which tend to be fixed and long term, include (1) the obvious and visible elements of hardware and equipment and the buildings to house them in, and (2) the people needed to operate, maintain, and support that physical plant. The former typically amounts to about 40 percent of corporate IS budgets, the latter 60 percent.

Senior management, if it takes any notice at all, is likely to be unhelpful. Top management response to the fixed-cost nature of supplier costs is usually to demand that IS budgets be cut. In late 1989, for example, top management in one of the world's twenty largest banks, viewing IS costs as too high, decided to cut the firm's IT budget from $520 million to $400 million. IS expenditures on maintenance, operations, and ongoing development represented $460 million of the $520 million, and as support for core business services such as credit card processing, automated teller machines, corporate electronic banking services, and basic processing systems could not easily be cut, IS's only practical choice was to cut corners on testing, operations, and security, which meant cutting quality, service, and reliability. The bank subsequently fired the hapless head of IS; his successor is likely gone by now as well.

It is easy for managers to demand that IT costs be brought under control, but it is reasonable for them to do so only if they acquaint themselves with the origins of those costs. They should understand, for example, that real expenditures grow much faster than software development budgets. We have seen that today's systems development costs compound tomorrow's committed budget for operations and maintenance.

Senior management should also recognize that infrastructure costs dominate applications costs. A firm that moves toward basing most or even much of its operations, product delivery, and customer service on its IT base has to make greater and greater investments in the infrastructures that support the practical and cost-efficient creation of individual applications. There is no cheap and easy way to build the global telecommunications networks and large-scale data and network management systems that constitute the corporate IT infrastructure.

Most senior business managers, and many IS managers as well, are unaware of these basic realities. Tradition has led them to view IT as an annual expenditure. In good times, the IS manager makes the case for an increase of X percent, or adds up all the approved new software development requests from the business units and factors in aggregate operations costs. In bad times, senior management demands either a decrease in the rate of growth or even an absolute cut in the budget (a very new phenomenon for IS).

Both approaches ignore the cost dynamics of IT. Figure 6-2 shows

FIGURE 6-2   The Cost Dynamics of Information Technology

Software development drives the cost base; typically, each dollar of development generates 20 cents of operations and 40 cents of maintenance.

In this example, development is 25% of the total budget in year 1 of $40 million.

Strategy 1: Maintain a Level Budget of $40 Million

Key:

Total budget ($ million)
Development budget
Total operations and maintenance budget

40.0	40.0	40.0	40.0	40.0
10.0	4.0	1.6	1.0	0.7
30.0	36.0	38.4	39.0	39.3
Year 1	Year 2	Year 3	Year 4	Year 5

Development has been cut by 93%, but the overall budget remains the same.

Strategy 2: Keep Development Level at $10 Million

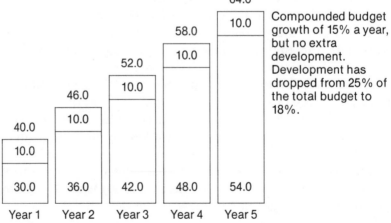

				64.0
			58.0	10.0
		52.0	10.0	
	46.0	10.0		
40.0	10.0			
10.0				
30.0	36.0	42.0	48.0	54.0
Year 1	Year 2	Year 3	Year 4	Year 5

Compounded budget growth of 15% a year, but no extra development. Development has dropped from 25% of the total budget to 18%.

FIGURE 6-2 Continued

Strategy 3: Grow Development by 10% a Year

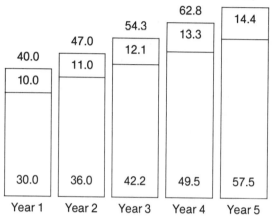

Compounded budget growth of 18% a year, for just 10% growth in development. Development now 20% of total budget.

Strategy 4: Grow Development by 20% a Year

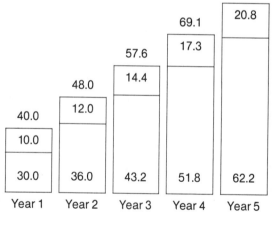

Keeps development at 25% of total budget and rate of growth of both development and total at 20%. This provides a steady balance between total budget growth and development growth. Many studies suggest that the natural rate of growth in demand for development is at least 20% a year. That means a budget of over $200 million in year 10.

how IT costs are compounded, assuming that every dollar of development generates 20 cents of operations and 40 cents of maintenance.

Following strategy 1, which calls for maintaining a level total budget given the reality of compounding IT costs, would all but eliminate development within five years, virtually assuring that the firm would suffer competitively either from skimping on maintenance and operations and thus on customer service, quality, and reliability, or on development and thus on business innovation. Maintaining or growing development, as called for by strategies 2–4, is not possible without budget growth. A mere 10 percent growth in development, as in strategy 3, would increase the overall IT budget by a factor of four in just five years.

Figure 6-2 oversimplifies the details of IT budgeting but not the reality of compounding costs and of systems development as the agent of compounding. In light of these figures, the 15 percent rate of growth in IT expenditures across the U.S. economy in the 1970s and 1980s no longer seems so large. It is clearly not enough even to keep systems development level to ensure adequate operations and maintenance. The natural rate of growth is closer to 20 percent. For firms that are aggressively spending on IT for effective competitive positioning, the rate is often closer to 30 percent. This latter rate obviously can be sustained for only a few years unless the benefit flow keeps pace with the costs. But unless a firm commits to a growth of at least 15–20 percent per year, it is effectively cutting back its IT capability unless it can dramatically improve development productivity, a challenge that the IT field has not yet made more than token progress in meeting. Software development remains a complex and slow craft.

Infrastructure investments compound the problem of managing IT costs by adding to the fixed-cost base without providing direct benefits. The development expenditures shown in Figure 6-2 can be assumed to have been justified by some business case showing economic returns. If the case is realistic and the costs fully stated—two extremely big "ifs"—then the firm will benefit and there is a rational basis for funding the development. Far more often, the critical investments are not for specific business applications but for infrastructures such as an international telecommunications system, customer information data base, or network management facility. Infrastructures are all cost; benefits come indirectly from the business applications they make practical.

Infrastructures can rarely be cost justified. As more and more development, and to a lesser extent operations, is distributed to business units, the central IS unit's main responsibility is for central systems and shared infrastructure resources, especially the corporate "backbone" network.

There are many opportunities to reduce infrastructure costs by consolidating separate facilities, especially in the area of international telecommunications. One firm, for example, cut its total telecommunications costs by 30 percent by linking separate services to a shared high-speed, low-unit-cost fiber optic transmission facility, centralizing network management, and routing international traffic so as to avoid high-cost countries. It should, of course, have done this years ago, but no one in the firm—literally—knew either how large the costs of telecommunications had become or where they were incurred; expenditures on telex, fax, international phone calls, outside data communications services, and internal facilities were scattered across several hundred budgets. It took several months of study to identify them.

The commonsense development of IT infrastructures is analogous to the commonsense upgrading of the infrastructures of congested, inefficient, and disruption-prone airports such as Heathrow, Kennedy, Frankfurt, and San Francisco, which is being totally blocked by politics, costs, and problems of location and transition. Many people complain that something should be done but are unwilling to pay any personal costs of money, effort, and inconvenience.

With costs growing far faster than business profits, evidence of payoff limited or absent, and small sanctioned increases in annual budgets generating long-term expenses, it is no wonder that corporate IS units are under siege. Frustrated and worried, senior management asks, "How on earth, with the costs of personal computers dropping by the week and the price/performance of the microelectronics industry continuing to improve by 30 percent per year, can our IS spending be growing by 10–20 percent per year with no increase in development or business productivity?"

The answer—"That's the way it is"—is not a justification for uncontrolled IS budget growth but a challenge to *business* managers to face up to the reality of IT costs and work to ensure that the returns from IT investments exceed those costs.

What Is Management to Do?

Not to allocate central IT costs is to provide a free resource, eliminating any sensible basis for rationing, avoiding waste, or controlling costs. Corporate IS clearly must charge its costs out, but finding a rational and acceptable way of doing so is difficult. Allocating costs on the basis of full cost recovery affords users little opportunity to manage their own costs. They simply incur an annual charge to their budget and may pay extra for increases in infrastructure investments from which they receive no immediate benefit. Additionally, the cost of costing—of tracking usage, billing, auditing, and so forth—can be enormous.[4]

Allocations give business units ample reason to be displeased with corporate IS. They see large charges at the same time they hear about rapidly falling costs of technology. They naturally wish to exploit the latter and avoid the former. So the cry goes up: "Break up corporate IS!" Or outsource it.

Someone needs to voice the countercry: "No! Rethink the management process for IT." There are no magic, plug-in solutions to the problem of managing IT costs, only commonsense directions for improving, if not resolving, the situation. Management must first ensure that the full lifecycle costs of IT, including all relevant hidden costs, are identified and included in any analysis of IT options.

Further, management must develop an asset, instead of a budget, view of these costs. This means funding the corporate infrastructure separately from business applications. The infrastructure is a shared corporate resource that enables a range of both foreseeable and unpredictable uses. Telecommunications networks, shared data base management resources, and security and network management utilities are all part of the infrastructure and should be funded by top corporate or divisional management as a long-range capital investment justified by corporate policy requirements.

Supplier costs can be recovered through a "quasi-profit center," eliminating cost-based allocations. This follows from the principle of funding IT infrastructures separately and allowing them to recover costs in a manner analogous to a power utility, whose managers spend capital and petition state regulators to allow them to set rates that recover the investment over a period of years at a "fair" rate of return. The utility sets prices; it does not allocate costs. The difference is key. Imagine going into McDonald's and being charged $227 for a Big Mac, the manager explaining that this is the fully allocated costs of the hamburger plus $1; he has just had the kitchen remodeled and must pass on the cost. IS has traditionally been required to allocate costs in this way. The quasi-profit center approach allows IS to set prices sensibly, for example, to price a new electronic mail service below cost to stimulate demand and make sure that early users do not have to pay the heavy initial installation and operations costs.

IS should not, except in special circumstances, become a real profit center. A true profit center, such as a separate IT services company, easily creates a business contradiction. If its services are outstanding it should be totally devoted to the needs of the business that owns it instead of being diverted to serve outside firms. Yet if it is expected to maximize its profits, it has little reason to support the firm's IT needs except where it can command a premium price. Only if it is mediocre should it be allowed to focus on external markets, in which case it makes no business sense for the firm to own it.

Many firms that have adopted the quasi-profit center approach base the pricing policy on a market basket of outside providers' prices. The manager of the telecommunications network, for instance, may be expected to offer services at 80 percent of the average of five specific vendors. Some will allow users to buy certain services outside, limiting them only in cases that affect overall corporate costs, security, and resource sharing. The internal IS organization is then part of a regulated free market. One major additional advantage of quasi-profit center pricing is that it reduces the cost of costing, tracking the details of costs, and establishing often complex allocation formulas.

To protect systems development, management must ensure that development proposals are screened by business managers, costs are fully addressed, benefits are fully and systematically assessed, and that the manager making the claim on the capital is held accountable for delivering the benefits. No IS manager can be expected to set the business priorities for a system and identify system benefits and guarantee their delivery.

Finally, management should encourage the reuse of existing resources. Much of the opportunity to use IT competitively derives from repackaging existing information, and many of the most promising long-term approaches to easing the software development bottleneck relate to reusing program code.

For management to take these initiatives, responsibility for IT application planning must be embedded in the business units. Information Services' responsibilities should lie in infrastructure planning at the corporate level, with senior business management setting policy and Information Services creating the needed architecture.

Pushing responsibility for IT application planning into the business end represents a major organizational shift for many firms. It directly links the issue of managing costs with managing benefits and reveals a question that remains unasked, never mind unanswered, in most firms: "Who is accountable for IT benefits?"

Summing Up: Making Sure to Get the Hidden Costs

Figure 6-3 lists apparent versus hidden costs for two components of IT expenditure, software development and purchased software packages. It shows how large a fraction of the full cost is incurred outside the activities of the programmers who write code or outside the purchase price of a package.

Software development costs vary by size and type of project, type of technology, quality of staff and project management, and number of development tools and techniques. Figure 6-3 presents the cost components of software for typical types of business applications that would constitute the bulk of an Information Services unit's portfolio. It shows how many of them are largely hidden and ignored in the business justification and budget. Few projects, for instance, budget twice as much for education as for programming the computer code.

Improving software development costs and productivity is a major problem for every large organization. It has been so since the 1960s and is likely to remain so through the 1990s despite efforts by vendors, computer scientists, researchers, and consultants. NASA, for example, carried out a project to develop "perfect" software, software that is close to error-free. It succeeded, at a cost of $1,000 per line of program code. The cost for a commercial organization is typically between $20 and $50 per line. NASA reported that it found no high-tech tools or new techniques that ensured the level of quality it sought; its approach was the tried and true "inspect, test, retest."

Computer-assisted software engineering (CASE) is a promising approach to improving development productivity. It uses computer tools to help computer staff, through a mix of design disciplines, analytic tools, and design and document management aids. CASE is part of the emerging organizational and technical infrastructure of IS. It is, however, roughly at the same stage as personal computers were a decade ago. It also demands major shifts in IS professionals' work, skills, attitudes, and effort; they often show as much "resistance to change" as their "users" did when IS brought them new systems.

FIGURE 6-3   **Apparent versus Hidden Costs of Software**

1) Percentage of Expected Project Costs for Development and Installation:

	For customized software develop- ment project			For purchased software package	
	Expected	Hidden		Expected	Hidden
Planning and Design	40%		Planning and Design		40%
Programming	10%		Analysis, Evaluation	25%	
Testing	10%	20%	Testing	5%	10%
Installation	10%	10%	Installation		20%
Total	70%	30%	Total	30%	70%

2) Follow-on Organizational Costs (as percentage of Total Project Costs)

	Expected	Hidden		Expected	Hidden
Education		20%	Education		20%
Support, Consulting			Support, Consulting		10%
Total		20%	Total		30%

3) Follow-on Lifecycle Costs (as percentage of Total Project Costs) Over First Four Years of Use

	Expected	Hidden		Expected	Hidden
Operations		80%	Operations		80%
Maintenance		160%	Maintenance		10%
Total		240%	Total		90%

4) Summary of Total Lifecycle Costs Over First Four Years

Actual Project Development Costs	100%	100%
Planned Project Development Costs	70%	30%
Actual Lifecycle Costs	360%	220%
Lifecycle Costs as Percentage of Expected Development Costs	514%	733%
Ratio of Expected to Actual Costs	1:4	1:7

(These figures are averages for medium-sized projects, with budgets of $1 million to $15 million; the source of the estimates is the author's research and consulting in four companies. The exact ratios and percentages will vary by company and project. If the ratio of those expected and hence budgeted costs to hidden costs is high, the entire project is off-track on day 1. So is the Information Services' unit's credibility.)

Packages and related "fourth-generation" development languages are often seen as one solution to the problem of software development productivity. Packages, which are general-purpose software that can be adapted to meet specific needs, substitute for the assignment of a team of corporate IS analysts and programmers to custom-build a system. Packaged software may not satisfy all the requirements of its users. IT people refer to this as the 80–20 rule; users get 80 percent of what they want for 20 percent of the cost. To get the other 20 percent they will likely have to spend an additional 80 percent.

The same is true for fourth-generation languages (4GLs), special-purpose development tools that can dramatically reduce the cost of developing small-scale and ad hoc applications. Personal computer spreadsheet software and many data base management, graphics, electronic publishing, and accounting systems are variants on packages and "4GLs."

The price of packages alone should not be the basis for making the decision to build or buy. A package may be a bargain at twice the cost of building a customized system if it reduces the maintenance burden. Conversely, it may be far more expensive in real terms at half the cost of building a system in-house when support and operations are included.

Almost every new increase in software productivity has been wildly overhyped when it first appeared. Fourth-generation languages were promised as the path to applications development without programmers, just as office automation was supposed to create a paperless office of the future in just a year. CASE was expected to generate automatic code. All these tools offer value but demand time, change, learning, and effort. There is no instant bargain.

That is also true for personal computers, which are widely seen as the best bargain in IT. But the term "personal computer" is in many ways misleading. The telephone is "personal," too, but it depends on a complex infrastructure, and telephone charges include an allocation of the costs for this infrastructure. To the extent that personal computer users want to connect their machines to others and access information stored on larger computers, they will need an equivalent infrastructure. In calculating the real costs of personal computer use in organizations several years ago, the Gartner Group likened the obvious cost to that for the telephone handset, and the hidden costs to those of the telephone company.

Gartner calculated that the hidden additional cost for a personal computer with software and printer, bought for just over $4,000, is close to $20,000.[5]

The organizational IT infrastructure requires an entirely different kind of business justification than do specific business applications and incidental hardware. The business case for support infrastructures must consider trade-offs between longer-term corporate needs and current application-specific needs. It involves paying a premium to invest in infrastructures that ensure continued flexibility and quality of service and that help the firm avoid being pushed into a position of competitive disadvantage. The decision to pay that premium is, like any other decision to make a strategic capital investment, part of the senior business-management policy agenda. The cost is likely to amount to as much as 10 percent of the corporate IT budget; in many large firms, these infrastructures amount to as much as 30 percent.

It generally takes about seven years to bring to completion any major business innovation that depends on building a comprehensive IT infrastructure such as a global telecommunications network or new generation of data base management systems.[6] This means that the components of the IT expenditure for infrastructures cannot be allocated as a direct cost but must be treated as a corporate capital investment. Though the cost of application development and operations can be charged directly to a business unit, to make the first users of an infrastructure bear the costs is to block innovation.

Managing IT Benefits

Managing the cost side of the IT economic equation is difficult. Managing the benefit side is a virtually intractable problem. There is no evidence that U.S. business's massive investments in IT over the past decade have significantly improved productivity or economic performance. Without such evidence, business managers are understandably skeptical of new promises and claims. Much of the problem derives from the historical treatment of IT as an overhead expense rather than a capital asset. IT, like most capital investments, involves long lead times, depends on often complex organizational learning, and requires parts of the overall infrastructure to be in place before specific applications can be built. In this context, trying to relate impact to expenditure is a meaningless exercise. Not only should we expect no correlation between 1985's IT spending and 1986's or even 1990's profits, we should anticipate a drain on profits during periods of investment in IT infrastructures.

We do not yet know the relevant time frame for assessing the impact of IT investments. It could be as much as fifty years. Richard Franke reports that

> during the British Industrial Revolution following the introduction of Watt's coal-fired steam engine in 1775, there were no early increases of output and efficiency. Instead, time seems to have been needed for technology diffusion and for human and organizational adjustment. It was not until the 1820s, about a half century later, that there began substantial increases of output, productivity, and income.[7]

Taking the long-term view does not help the top management team concerned with making practical policy decisions today to allocate scarce capital to IT. And the problem cannot be finessed by playing with accounting system ratios and formulas. There is simply no reliable way to measure the value-added benefits of IT.

Here again, history suggests that IT is not a unique technology. Economists still debate the impacts of the railroads on U.S. economic performance in the nineteenth and early twentieth centuries. Common sense tells us the impact was huge; the figures do not. Many of the benefits came from rethinking old ways of operating and from new services. One major study concludes that the railroads contributed to an initial 1.5 percent increase in productivity, which rose to 4.5 percent a year after several decades, through value-added uses of railroad infrastructures such as improved freight scheduling.[8]

IT, like the railroads, is primarily an enabling technology. Much of its impact will be manifested in creative uses of technologies rather than in the technologies themselves. These impacts may be hard to link directly to IT.

Just as there is no reliable way to measure the benefits of IT, there is no reliable yardstick for deciding how much to spend on it. Many senior managers calibrate their own firms' IT spending with industry averages, typically X percent of revenues, and use that as a gauge to judge whether costs are too high or too low. Such ratios are effectively meaningless. If the industry average is 2.4 percent of sales and a firm is spending 8 percent, is that firm a breakaway company, or is it spending stupidly? Is it funding infrastructures that others are neglecting, or is it catching up after years of underspending?

The problem of measuring the business value of IT will be a core management issue over the next decade. The current context is one of growing costs and limited or no evidence of payoff. The business case for investing in IT is straightforward: does it provide a higher rate of return than putting the money into some other area of the business? Business logic, not financial evidence, will determine whether the infrastructures that create the firm's technical platform are essential capital investments.

What is most worrying about the current situation is the pressure to take a very short-term view of IT. The growing reliance on outsourcing and facilities management services for providing IT appears to be driven by management's desire to reduce debt on the balance sheet, whether as a defensive measure against takeovers, to leverage short-term share price, or to position for a management buyout. Getting IT under control and out of the firm can contribute, but at what longer-term cost? Are we witnessing another instance of expediency that mortgages business's future? U.S. neglect of other long-term investments in manufacturing and R&D is a warning signal; neglect of longer-term IT investments is already the norm rather than the exception.

Donald Trump justified his effort to acquire American Airlines on the basis that the airline was not passing profits back to shareholders, who in his view were being poorly served. American has led its industry over the past decade in every measure of marketplace performance: revenue and profit growth, business innovation, fleet modernization, capacity expansion, and information technology. It is one of the outstanding U.S. competitors in worldwide industry, having consistently invested for the long term to ensure a flow of profits. The Trump mentality would almost certainly not have generated Sabre, the IT platform that has contributed so much to American Airlines' preeminence. All the problems and dilemmas of IT costs, unproven benefits, risks, and long lead times that are discussed in this chapter make it an easy target for gutting instead of building.

Assessing the Business Value of IT

IT benefits can sometimes be measured convincingly at the application level. An electronic mail system, for example, can be evaluated in quantitative terms by tracking the costs of communication. A competitive initiative, such as a new product, can similarly be evaluated through changes in market share, direct revenues, and customer satisfaction. It may be hard to place an exact dollar value on these benefits, but they can at least be systematically assessed.

Many a scholar, consultant, and practitioner has tried to devise a reliable approach to measuring the business value of IT at the level of the firm. None has succeeded.[9] Given the difficulty economists have had trying to measure the economic impact and value of the railroads in the nineteenth century, of innovations in ship technology and design, and of the creation of electric utilities, there is little reason to believe that we will solve the conceptual and methodological problems associated with measuring the value of IT within the next decade.

Yet doing so is a growing management concern, not an academic exercise. Senior business executives need answers to some simple questions. "What is the basis for deciding if we are spending too much or too little on IT?," "How do we tell if we are getting real value from IT?," and "How do we monitor competitors' IT expenditures and benefits?"

Today, managers see mainly the costs of IT. They have heard claims and promises of its benefits, and they have seen some of these, albeit seldom in dollar terms. If IT were a static fraction of the overhead budget, they might be content to leave it at that. But with IT costs now half of a firm's total capital expenditure, that is no longer an option.

A number of practical principles for managing the benefits of IT are emerging. The operational term here is "managing" as opposed to "measuring." The history of IT has not seen the development of a standard financial yardstick for assessing IT value. Paul Strassman has shown convincingly that industry IT budget-to-sales ratios, profit comparisons, and analyses of return on assets show no correlation between firm performance and IT investment.[10] That should not be surprising.

First, IT spending does not create benefits, any more than R&D spending does. Throwing money at technology makes about as much sense as throwing money at R&D. As with R&D, benefits often lag, and are sometimes not directly linked to the immediate and highly visible costs that give rise to them. Imagine trying to measure the business value of management education. Most companies accept the need to invest in education and do not expect today's education budget to yield a tidy ROI figure tomorrow. Neither are they very likely five years from now to be able to measure a direct payoff from that education expenditure.

Second, the same investment in the same technology can have

very different outcomes. It is the management process rather than the technology that determines the benefits. In a given industry every firm will have access to the same base technology.

Third, many of the value-added benefits of IT do not show up in the accounting system. IT benefits, like education benefits, do not translate tidily into cost savings, especially when the investment is made in order to enhance service, effectiveness, customer image, and that mysterious element "productivity." Furthermore, standard financial measures distort the value of IT that is managed as budgeted overhead. No one can measure the business value of overhead. Just as the accounting system hides many aspects of the costs of IT, it overlooks many of the ways in which IT contributes to business units' performance. The most important contribution IT can make is in avoiding costs. For example, by adding IT costs in year X, a firm might avoid increased business costs in year X plus 5. Looking just at financial records, the business manager sees a growth in IT expenditure and a reduction in business cost growth. IT is now 12 percent of the company's total cost base versus 8 percent previously, while the business units have managed to reduce labor costs as a percentage of sales by 30 percent. What's going on? The business manager's likely conclusion: "IT is out of control!" But what *may* be going on is that the IT investment generated the labor cost savings by ensuring that the firm could handle larger volumes without adding staff. The accounting system will never tell!

Finally, the IT platform is the infrastructure for the later benefit flows. A firm's telecommunications infrastructure, which shows up only as a cost, makes practical many services and products that would otherwise be infeasible or prohibitively expensive. Benefits are attributed to the product, and costs to the infrastructure. It makes sense to charge a standard usage fee for a mature infrastructure such as an electrical utility, a toll road, or a public telephone system. Measuring its business value and the value of reinvesting in it to avoid obsolescence and exploit new technological developments depends on being able to relate original costs to the later flow of benefits minus recovered costs. That demands far more sophisticated planning and accounting systems than most firms use for IT.

The most obvious interpretation of Figure 6-4, which summarizes Strassman's empirical findings regarding the relationship between IT expenditures and measures of business performance, is that IT does *not* provide real benefits.[11] This is the IT equivalent of the story of the drunkard who searches for his lost keys under a lamppost

FIGURE 6-4 **Technology and Economic Payback**

Paul Strassman has provided the most comprehensive analysis of studies of the economic payoff from IT investments. Nathan Rosenberg has similarly summarized research on the more general relationship between technology and productivity.

Strassman's Analyses of the Payoff from Information Technology

Industries that invested most heavily in IT have shown relatively poor productivity growth;

No correlation between a firm's return on assets or return on investment and its spending on IT;

If anything, heavy computer users show a poorer, not stronger, ROA than average;

Study after study indicates firms' lack of ability to measure the business value of IT and continued frustration in trying to do so;

No proven cost-benefit techniques;

Comparisons of spending levels as a percentage of revenues or cost base meaningless; firms within the same industry vary too widely in such areas as sales per employee, ROI, asset base, etc.;

High or low IT expenditures give no indication of the effectiveness of spending;

Presently available methods are inadequate to satisfy management needs; senior executives remain frustrated with the problem of assessing IT;

The most profitable consequences of IT investment appear to come from long-term restructuring and simplification of internal and external communications.

Rosenberg's Summary of Research in Economics on the Payoffs from Technology

Economists have wrestled throughout this century with quantifying the impacts of the railroads and ship technology on productivity. In the case of ships, it is almost impossible to pin down the impact, even though two centuries saw a continuing flow of technical improvements and a continuing flow of cost improvements in freight; the cause and effect relationships are obscure and indirect.

"Really major improvements in productivity . . . seldom flow from single technological innovations, however significant they may appear to be. But the combined effects of large numbers of improvements within a technological system may be immense."

The impacts of a specific technology often depend on second-order developments; for instance, until the invention of the meter, supply of electricity depended on fixed-cost contracts to large customers since there was no way to measure individual usage. The meter expanded the market and created the utilities we take for granted today.

The impact of many technologies depends on an accumulation of small improvements; "Most of the developments in general-purpose digital computers resulted from small, undetectable improvements, but when they were combined they produced the fantastic advances that have occurred since 1940."

FIGURE 6-4 **Continued**

Rosenberg's Summary of Research in Economics on the Payoffs from Technology

The cost reductions in business that result from the "alpha" phase of a new technology are often far smaller than later benefits from the "beta" phase; "The evidence from the petroleum refining industry in-

dicates that improving a process contributes even more to technological progress than does its initial development."

"Learning by using" explains much of the productivity, rather than just the availability of the technology. This learning process may take decades.

that is nowhere near where he dropped them, because it provides light for him to look.

One reason to stop looking for IT productivity benefits under the financial reporting lamppost is that IT rarely reduces costs. Its main value is more often in changing the cost structure of the firm so that it can increase volumes without increasing personnel. This is shown in Figure 6-5, which shows how IT substitutes fixed-cost capital for variable-cost labor. In many service- and information-intensive areas of business, this offers an opportunity for major cost avoidance. Such activities are typically marked by diseconomies of scale. As volumes increase, administration and coordination grow faster. It is a frequent cause of the organizational complexity discussed earlier. Panel (b) in Figure 6-5 shows the result: decreasing margins if costs rise to meet service and administration needs, or decreasing quality and responsiveness if they are kept down.

Panel (c) in Figure 6-5 shows the impacts of effective use of IT. Fixed costs increase but the rate of growth in variable costs is slowed. The gap between the two cost trends is a real saving, directly attributable to IT. However, company costs have still increased. Indeed, the IT investment may have contributed to the increase by allowing the company to take on volumes of business that would otherwise have been infeasible. Airlines, banks, and credit card firms could not possibly operate at today's levels of activity without their IT base; their labor costs would be stratospheric and the inevitable errors and delays would result in a totally unacceptable level of service.

Figure 6-5 also summarizes the major impact of IT on costs: avoidance rather than reduction. It also points up a major dilemma in making the business case and measuring the resulting business

FIGURE 6-5 **Productivity and Information Technology**

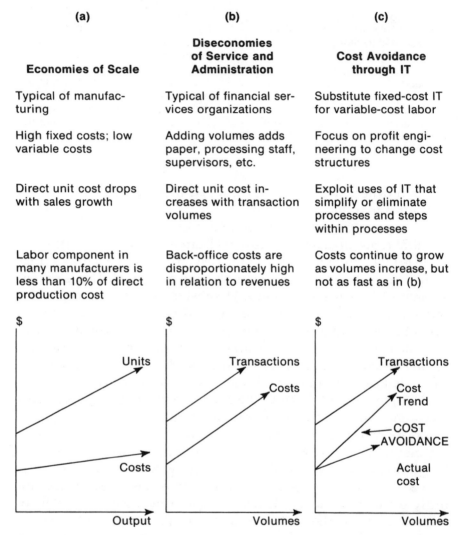

(a)	(b)	(c)
Economies of Scale	**Diseconomies of Service and Administration**	**Cost Avoidance through IT**
Typical of manufacturing	Typical of financial services organizations	Substitute fixed-cost IT for variable-cost labor
High fixed costs; low variable costs	Adding volumes adds paper, processing staff, supervisors, etc.	Focus on profit engineering to change cost structures
Direct unit cost drops with sales growth	Direct unit cost increases with transaction volumes	Exploit uses of IT that simplify or eliminate processes and steps within processes
Labor component in many manufacturers is less than 10% of direct production cost	Back-office costs are disproportionately high in relation to revenues	Costs continue to grow as volumes increase, but not as fast as in (b)

value of IT. Costs still go up and the promised cost avoidance too often does not occur. The time gap between the initial investment and the benefits may be very long, as shown in panel (c) of Figure 6-5. It may be hard to isolate the effects of IT in this context.

Establishing Anchor Measures

The preceding example points to the need to ensure that IT investments are assessed in relation to an *anchor measure*, an operational indicator of performance that can be used over time to identify overall IT impact. In Figure 6-6, panel (c), there is no anchor; the firm's costs continue to rise, and the natural interpretation is that IT is not providing a real payoff. Yet the administration and service costs per transaction are obviously declining. By that anchor measure, IT has generated measurable benefits.

The appropriateness of the anchor measure is a management judgment. No measure is "correct." The choice guides investment and represents a statement about how to judge the impact of IT. The anchor measures that many companies have adopted by default often limit IT investment in infrastructures; examples are standard accounting measures such as cost savings, comparison with industry IT spending as a percentage of sales, and analysis of IT budgets and the firm's return on assets. These have certainly been effective in the sense that they have guided IT decisions, even if not in a sensible direction.

For a retail bank, likely *appropriate* anchor measures are (if the major thrust in the firm's use of IT is to improve its cost dynamics) the annual cost of serving a customer or costs per transaction. These would not be appropriate anchors if the priority were market and revenue growth. Then, suitable anchors might be revenue or products per customer. In manufacturing, the anchors for an electronic data interchange investment might be document costs per shipment or average delivery time in days.

Anchor measures may not be readily translatable into quantitative financial figures such as return on investment. That is the wrong lamppost. The anchors provide the equivalent of a Dow Jones average or Consumer Price Index that answers the question, "How are we doing?" The Dow Jones average is not an exact measure of stock market movements; the mix of companies whose share price is used to calculate the index is not fully representative of the market. Similarly, the CPI is oversensitive to fluctuations in house prices. The indices are not precise measures of the stock market or of inflation, merely operational indicators of general direction and progress.

The typical financial measures firms use will rarely be suitable anchors for assessing IT progress and impact. Most profitability and return on asset figures penalize IT investments, since the development investment in year 1 increases the expenditure base and later the asset base (the hardware is added to fixed assets) but the benefit stream may not originate until year 7. That is an obvious feature of capital investments in general; no one expects a new office building, research and development, or education program to generate imme-

FIGURE 6-6   **Anchor Measures of Profit for IT**

Traditional profit measures largely penalize IT investments, as this example shows. It is based on an actual financial services organization.

(a) extrapolates the company's performance over a three-year period, beginning in 1984, applying historical trends in costs, sales, and so on. The figures for 1987 are extrapolations.
(b) shows the actual figures for 1987, following a massive investment in IT.
(c) and (d) show the difference between the two when return on assets is used as a yardstick versus the recommended anchor measure of revenues and profits per employee.

	(a)			(b)		
	Actual for 1984	1987 estimate based on past trends	Increase/ decrease (%)	Actual 1987 figures	Increase/ decrease (%)	Actual looks better than trend
Assets ($bn)	2.1	2.3	9.5	3.3	42.8	?
Sales ($bn)	1.2	1.6	33.3	1.8	33.3	Yes
Profits ($mn)	60	72	20.0	84	40.0	Yes
Labor ($mn)	404	560	38.6	161	−60.0	Yes
IT ($mn)	102	124	21.6	182	78.4	No
Employees (thousands)	15.3	19.4	26.7	6.9	−54.5	Yes
Return on assets (%)	2.8	3.1	10.7	2.5	−7.9	No
Sales per employee ($ thousands)	78	82	5.1	261	234.6	Yes
Profits per employee ($ thousands)	3.9	3.7	5.4	12.2	303.1	Yes
Labor costs as % of sales	33.7	35.0	3.9	8.9	−164.1	Yes
IT costs as % of sales	8.5	7.8	−10.7	10.1	18.8	No
IT costs as % of labor $	25.2	22.1	−12.3	113.0	348.4	No

Interpretations:
1) IT is out of control! The business units have cut their costs but look at IT expenditures in 1987.
2) What a much more healthy firm this is in 1987 versus 1984. What a great contribution IT has made.
3) If we used return on assets as the yardstick, we would not have made the IT investments.
4) If we had made return on employees the yardstick, we would have started our IT drive earlier.

(c)
Mapping ROA and profits

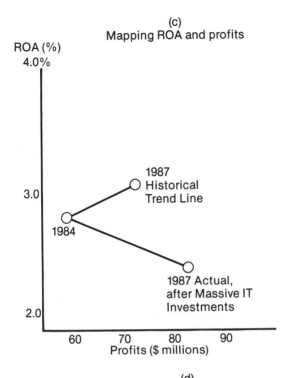

(d)
Using sales/profits
per employee as the anchor
measure for assessing IT

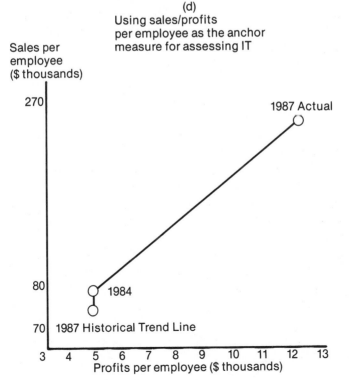

diate payoff. Treating IT as an expenditure instead of an asset has created such an unrealistic expectation and encouraged a very short-term focus among senior managers.

The impact of choice of anchor measures, either explicitly or by default, for assessing IT's contribution to the overall firm performance is shown in Figure 6-6. It shows the performance of a firm in the financial services industry between 1984 and 1987. Panel (a) extrapolates from historical trends. Panel (b) shows the actual results. The other two panels interpret those results. Panel (c) shows that the firm's investment in IT was a disaster. It increased the asset base from 2.1 billion dollars to 3.3. Return on assets dropped. IT costs went up by close to 80 percent. The business units cut staff from 15,000 to just under 7,000.

As panel (d) indicates, however, IT may have saved this firm. Aggressive investments in IT cut staff by over 200 percent and leveraged their contribution to the firm's profits by 300 percent. This company is one of the leaders in the United States in using information technology to speed up service and improve marketing. Yet panel (c) shows that IT has been a drain on performance.

Panel (d) shows the opposite and is a far more appropriate anchor measure for assessing IT. It is appropriate, however, only if management buys in to the following business priorities and IT potential. All large firms must change their cost structures by reducing unit costs, eliminating unnecessary jobs and staff, and cutting out organizational layers without sacrificing customer service and quality. Pervasive investments in IT—e.g., in electronic data interchange, computer-integrated manufacturing, and telecommunications—can substitute technology for people in all the above areas, afford major longer-term opportunities to avoid costs, and help to generate additional revenues without proportionately increasing staff costs. In this context, which is very typical, the firm's cost structure shifts away from dependence on variable-cost labor to an increasingly fixed-cost technology base.

Tracking IT progress over time is most meaningful in relation to performance per employee. Are we exploiting IT mainly to gain leverage through organizational simplicity, zero-based work, and transformation of cost structures? Tracking revenues and profits will not provide a clear answer to these questions. Figure 6-6 (d) plots two dimensions: revenue per employee and profit per employee. Any firm should naturally aim at being so far along the forty-five degree line as it can. It does not want to move too far above the line, which indicates buying market share: high revenues, low profits.

Equally, it needs market growth and should not fall too far below the forty-five degree slope: low revenues, high profits.

Figure 6-6 is a useful managerial basis for planning and tracking IT investment over time. IT infrastructures are business enablers. The telecommunications infrastructure enables new products and services. Office technology networks enable new ways of coordinating teams. Data base management systems enable targeted marketing. EDI reduces errors and delays in paperwork. The management process, education, employee skills, choice of products, and innovation in marketing create the economic results. We cannot measure directly the impact of IT in many instances. We cannot reliably estimate the value of management education, another enabler. It is impossible to measure the value of overhead, such as photocopying.

The most obviously relevant anchor indicators of business-unit performance for assessing IT opportunities and progress are revenue/profit growth, leverage of staff, and the balance between short-term profits and investments that generate long-term new profits. Unless revenues and profits are tracked per employee, as shown in Figure 6-6 (d), there is little incentive for business managers to invest fixed-cost capital in IT. As an enabler, IT should, over a period of time, contribute to business performance in a combination of revenue and cost-related ways; otherwise the only reason for investing in it is absolute necessity, created by government and legal statutes or customer demand. In many instances, IT spending is driven by such necessity, but discretionary investments rest on judging where and how IT can:

- directly enable revenue growth without reducing profit margins;
- enable cost avoidance that will increase costs but decrease cost per unit of activity and hence cost per employee;
- leverage "productivity," which must be defined in terms of an operational anchor measure to avoid being just a vague buzzword (this again may increase the cost base but decrease cost per employee and may indirectly enable revenues per employee); and
- position the firm for future profits.

In every instance, absolute costs are likely to increase. The greater the investment in IT, the larger a fraction of the overall cost base it becomes. Crediting business units and not IS for the revenue or cost gains enabled by IT makes it look even more out of control. It is as impractical to try to measure the value of enabling infrastructures as it is to try to measure the business value of overhead.

Of course, when IT can directly and significantly reduce costs, there is no problem of measuring business value, and return on assets and investment ratios become more relevant. The anchor measure is simply cost displacement.

Obviously, any approach to measuring benefits depends on accurate costing. Firms must be sure the costs really are displaced. If hidden costs are ignored, the economic case is flawed from the start. The impact will become apparent over time, as later improvements in the firm's cost base (hence its profits) claimed as business justifications for specific IT application development projects never materialize.

Even when costs are accurately stated and the intended benefits carefully and realistically identified, the business justification is meaningless unless the business units agree that the benefit targets are realizable and they are held accountable for them. IT planners should be held accountable for supplier costs, but they cannot ensure that the IT enablers generate the planned results.

The need to make IT users accountable for IT benefits is illustrated by the case of Trident Bank (a pseudonym), the first in its country to move aggressively into using ATMs (automated teller machines) and telecommunications to deliver its retail banking services. The bank spent hundreds of millions of dollars on its ATM network. The ATMs generated revenues of $70 million, including fees from other banks whose customers used Trident's system. Operating costs were close to $55 million. The bank's staff costs were increasing, and its return on assets, profit margins, and revenue growth were average for the industry. Many business executives questioned the payoff.

It took little effort to show that the ATMs had provided a massive payoff. The $15 million gross profit margin ignored the 120 percent increase in transaction volumes over the past five years. Comparing the cost of handling a savings deposit, making a name and address change, and cashing a check via the IT delivery base versus through a teller showed that the bank had avoided costs of close to $200 million a year. Plotting the number of customer accounts and transactions handled per teller revealed that even though staff numbers and labor costs continued to grow, the ATMs were contributing to sustained and significant improvements in revenues and costs per employee.

The bank's profits in the year the study was made were just over $300 million. The ATM network had contributed at least $150 million through cost avoidance. An obvious question is why this had not shown up on the bottom line. Part of the answer is that the ATMs were cross-subsidizing other areas of operations. The number of accounts per *manager* was dropping. The number of transactions per *supervisor* similarly continued to fall. These managers and supervisors were not handling more complicated services than before, nor had the bank added a wide array of new products.

The benefits gained from improved teller productivity were lost because the bank had not taken the tough decision to reduce first-line managers. Moreover, the bank's executives tended to regard ATM benefits as the responsibility of the Information Services group. The apparent lack of acceptable payoff from the ATMs was attributed to IS.

Analysis of the bank's cash management service showed the opposite picture. What appeared to be a success was in fact a drain on profits. The direct development costs had been low and direct profit margins high. There were, however, large hidden costs of operations and an ongoing need to add expensive telecommunications facilities to handle growing numbers of customers who used the service only occasionally.

Who was taking charge of change here? The implementers of the new system had fallen into the drift described in Chapter 5; they added automation to existing jobs without rethinking work and work flows. They had deployed technology capital without adequately redeploying human capital.

The Trident scenario is fairly typical. Management is concerned that the company is not getting adequate payback from its heavy commitments to IT. IS costs are uniformly viewed as too high. Detailed analysis shows large cost avoidance that translates directly to payoff. That payoff is dissipated because the business units are not held responsible for ensuring the maximum practical level of benefits. The firm's accounting systems obscure relevant costs, and its measure of financial performance hides the impacts and relevance of IT.

Trident's management, though still concerned about IT costs, has learned many lessons. It is moving to make business units responsible for business justification and for delivery of the benefits promised by the justification. It is separating the funding of infrastructures from applications. And it is looking at overall longer-term costs of doing business, adjusting for volume changes, and rapidly eliminating mid-level staff and bureaucracy.

Managing IT Risk Exposure

When 25–80 percent of a firm's cash flow is on line, technology risk becomes business risk. Many business managers are unaware of the range of IT-related risks to which the firms are exposed. They learn about most of them only through disaster. For example, when a Bell Illinois switching center outside Chicago burned in 1988, many companies had no backup facilities. They do now.

Managers often have no idea how little protection their insurance coverage provides them. It protects them from damage to equipment, fraud, and theft. There is as yet no established insurance underwriting market to provide coverage for consequential losses of services caused by, say, software bugs, computer viruses, telecommunications transmission errors, or delays in development projects. Firms have growing risk exposure and limited risk coverage.

Two major areas of exposure are security and network management. Security is a growing problem in terms of vulnerability to criminal theft and fraud and to accidents of leakage of information. It is clearly difficult to combine access and control. The very idea of on-line customer service and product delivery is to make access convenient and easy for more and more people. Control demands the opposite: restriction and difficulty of access.

A study by a leading accounting firm of 35 leading companies' telecommunications strategies reported that 18 viewed lack of security on public data networks in Europe as a growing and urgent problem. Four have subsequently abandoned major business initiatives there because they viewed their risk exposure as unacceptable.

A 1986 survey of 100 U.S. accountants and 90 mid-level to senior-level IT professionals attending a conference on computer security found that three-quarters believed that most electronic thieves are caught by accident. "This is a startling admission of the vulnerability of the accounting controls, audit trails and programming documentation for which their professions are responsible."[12]

No one knows the true level of computer crimes—successful crimes may avoid detection entirely—but every study of those that are detected has found that they involve far greater sums of money than other white-collar crimes (excepting large-scale stock frauds and insider trading). It is far more expensive to build an IT infrastructure that maximizes security than to cross fingers and hope. If the IT infrastructure is a capital asset, firms will spend the money. If IT is an overhead expense, they will not. When most elements of core business operations depend on IT, it exposes them to high business risks.

Network management involves a different form of risk exposure that requires a highly complex technical infrastructure to protect the firm. A large-scale electronic services delivery system is extremely complex. A major network may link 50,000 personal computers, word processors, workstations, and telephones made by 50 different manufacturers to 500 computers supplied by somewhat fewer vendors across three to twelve time zones. Performance demands require 99.9 percent reliability. When the point-of-sale or computer-integrated manufacturing network is out of service, so is the business.

How much should a firm pay to minimize risk exposure in the areas of security and network management? It can pay a premium for various tools and sources of expertise. Security will be improved by encryption of telecommunications links, multilevel passwords for accessing computer systems, and diagnostic and auditing systems. Automated network management hardware and software can provide many facilities, including collection and display of network alarms, diagnostics, trouble ticket issuance, statistical reports in real-time, and network status monitoring.

All these facilities are expensive. Are they optional or essential? Automated network management systems add significantly to the cost and complexity of the computing and telecommunications base. Security and network management software systems are hard to add to individual applications; increasingly, they must be a component of the integral design of the technical architecture in order to be effective.

Viewing such facilities as an extra cost, and refusing to incur that cost, can put a firm at risk. Managing IT risk exposure is part of managing the economics of information capital. Just as many IT costs are hidden, so is much of its risk. Only a few elements of IT risk exposure can be laid off through insurance. Most can be reduced only by extra investment.

Building the IT asset balance sheet often jolts senior business managers into a new awareness of the level of investment IT represents. Analyzing IT-related risk exposure can put them in a stupor. Firms face constant, substantial, and multiple sources of risk associated with IT.

Equipment and facilities get most of the attention, but they can be insured through the property and casualty market. However, premiums and deductibles are increasing substantially, and coverage does not include loss of services. Reducing exposure means investing in expertise, backup facilities, and management tools. Network services, which represent a growing part of companies' cash flows, investing in automated network management systems, and recruiting and retraining people proficient in these systems are essential. Software errors can generate expensive business losses, as American Airlines' $50 million loss of revenue from an undetected bug in its yield management system demonstrated. Tools that can improve the quality of software will reduce risk, but at a cost. Finally, most software development projects depend on a handful of key staff; lose them and the project is essentially dead.

These are just some of the main sources of IT risk exposure. In-depth analysis of all components of the IT resource, including machines, software, facilities, suppliers, and labor, will inevitably reveal many others. Protecting the firm against these risks is the job not just of the IS manager but of the chief financial officer and controller. Reducing exposure almost surely means increasing costs.

The subject of this chapter has been one that every manager understands: money. IT is just one way a firm can spend its money to get business value. Managing IT is the same as managing money. Any manager who prides himself or herself on managing money should be just as proud of helping manage the money side of IT.

Notes

1. The figures provided throughout this chapter come from the author's work in a number of companies, mainly in financial services, airlines, and pharmaceutical firms. Obviously, they will not be the same across all firms. Most organizations are not in a position to challenge the ones given here, however, since they do not track the relevant information. Most Information Systems professionals also have little background on the practical economics of IT; a review of over thirty widely used IS textbooks in preparing this book revealed not a single chapter in any of them on the topics of the cost dynamics of IT. A similar analysis of over a hundred books and articles on the topic of information technology and competitive advantage shows not a single discussion of the cost base for the many exemplars shown. Most analyses of the economics of IT either focus on the IT industry itself or

use very abstract mathematical modeling techniques. The lack of sophistication, awareness, and practicality about the cost dynamics of IT among business and IS professionals is surely a stumbling block to progress and dialogue and must be high on any company's agenda for education and action.

2. Many IS managers will argue with the estimate of maintenance consuming half of an IS group's human resources, mainly because the definition of maintenance is often very loose. Some firms, for instance, only include repairs to programs and changes needed because of new operating systems, equipment, and so forth. They exclude such projects as changing a payroll system to handle new federal tax laws. These, however, are nondiscretionary demands that do not add new functionality but maintain the payroll system's usefulness and validity.

Other managers will disagree with the 50 percent figure because their organization has made a sustained drive to reduce maintenance costs by improving design and testing and by applying development productivity tools.

Very few IS managers are likely to argue with the estimate that only 10 percent of their staff are working on really new development rather than enhancement, redevelopment, and maintenance.

3. Much of this support is for education and troubleshooting. The term "user-friendly" is a misnomer; computers are less user-vicious than they once were but remain very hard to use. A typical pattern is that novices learn to use a few simple tools without understanding the details of the personal computer hardware and operating system; they then try out new ones and find that the details are important as they struggle to select telecommunications parameters for electronic mail, or specify the configuration for a new printer.

Support costs are likely to increase as a fraction of the purchase price of personal computers, not decrease. A fully loaded PC with powerful software, data base management, communications, and desktop publishing attributes, and linked to a local area network, is in reality a computer center more complex than the typical corporate center of the 1970s. Many of the hybrid career roles described in Chapter 5 are for support of IT users and uses.

4. A cogent argument for moving away from cost allocations and chargeback systems toward an internal quasi-profit center pricing is provided by Brandt Allen, "Make Information Services Pay Its Way," *Harvard Business Review* (January–February 1987), pp. 57–63.

5. See P.G.W. Keen and L.A. Woodman, "What to Do with All Those Micros," *Harvard Business Review* (September–October 1984), pp. 142–150.

6. For examples that support this estimate, see P.G.W. Keen, *Competing in Time* (Cambridge, MA: Ballinger, 1988).

7. Richard H. Franke, "Technology Revolution and Productivity Decline: The Case of US Banks," *Technology Forecasting and Social Change*, vol. 31 (1987), p. 143.

8. See N. Rosenberg, *Inside the Black Box: Technology and Economics* (Cambridge, England: Cambridge University Press, 1982).

9. For a detailed analysis of the efforts to show the links between IT expenditures and economic payoff, see Paul A. Strassman, "Management Productivity as an IT Measure," in *Measuring Business Value of Information Technologies* (Washington, DC: ICIT Press, 1988). Peter Weill showed that unless a "management conversion effectiveness" factor is included in the analysis of the relationship between IT investment and firm performance, studies largely show that, if anything, high spenders perform less well than low spenders. Conversion effectiveness is a measure of the quality of management and firmwide commitment to IT. See Peter Weill, *Do Computers Pay Off?* (Washington, DC: ICIT Press, 1991).

Strassman argues that IT expenditures are effective only if they improve management value-added and provide a detailed strategy for targeting investment. Both he and Weill demonstrate that spending money on IT is not in any way the same as getting business value from it.

10. Strassman, "Management Productivity as an IT Measure."

11. Ibid.

12. K. Hearnden, "Computer Criminals Are Human, Too," *Long Range Planning,* vol. 19(5) (October 1986), pp. 414–426.

Chapter 7

Positioning the IT
Platform

How can senior executives be sure that their company has an effective and appropriate technical strategy? Where will technical decisions most directly influence a firm's range of business options in the 1990s, either opening up potential opportunities or closing them down? What policy requirements does top management need to guide technical planning? This chapter addresses the *business* policy decisions needed for effective positioning of the information technology platform. It covers three main topics: the reach and range of a corporate IT platform; the top-level business policy questions that determine a firm's choice of technical strategy; and issues that face planners in developing an IT platform. The chapter contains a checklist of senior management policy criteria for establishing a technology base that will ensure adequate business flexibility and responsiveness.

The Reach and Range of the IT Platform

The business functionality of the corporate IT platform can be defined in terms of *reach* and *range* (see Figure 7-1). **Reach** refers to the locations a platform is capable of linking. The ideal is to be able to connect to anyone, anywhere, just as the phone system reaches across the world. This is as yet an impractical short-term target, but one that will become more realistic in the coming decade as standards allow different systems to interconnect much as countries'

FIGURE 7-1   **The Reach and Range of the IT Platform**

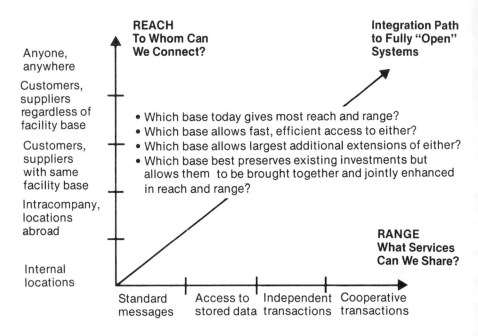

different telephone systems do today.[1] *Range* refers to the degree to which information can be *directly* and *automatically* shared across systems and services. The ideal here is that any computer-generated transaction, document, message, and even telephone call is able to be used in any other system, regardless of hardware or software base. Today, with widespread differences in proprietary (vendor-specific) software, operating systems, and information storage and management techniques, only systems built on the same hardware and software base can automatically and directly share messages, transactions, and data.

The alternative to a platform-centered approach, a case-by-case approach in which each major system has its own technology base, is supported by a number of practical arguments, among them the need to match technology to specific business, technical, and economic characteristics of applications, and the often immense difficulties involved in integrating separate IT resources. Today's operational needs often favor the case-by-case approach to IT investments. There are nevertheless very strong, longer-term business reasons for shifting from a piecemeal to a platform approach. Primary among these is the business degrees of freedom the latter affords a firm in using IT as a competitive and organizational resource.

Technical and Political Limits on Reach and Range

The goal of IT integration within and across firms—of hardware, software, information, and telecommunications working together as a single, integrated resource—is still a long way off. Many firms have no IT platform because information technology remains dominated by incompatibility between vendors' products and no one supplier can meet all or most needs equally cost-effectively. Firms that consequently operate a hundred or more separate business services on equipment provided by ten or twenty different vendors face major technical and cost barriers and many technical uncertainties as they try to balance conflicting demands for efficiency now and integration tomorrow.

Incompatibility—differences in vendors' hardware and software and in telecommunications facilities that prevent commonsense linkages—is the bane of integration. Today's typical personal computers provide a ready example. Data files created by a word processing or a spreadsheet program logically ought to be transferable to a data base management system running on the same machine. A new printer logically should not need a different type of cable than the one it replaces. A version of a particular software package that runs on vendor A's hardware logically ought to run on vendor B's—it's the same software. It should be easy to send and receive electronic mail messages to and from other PCs in the firm. In practice, "logically ought to" often means "doesn't."

The main source of incompatibility in IT is the operating system, the set of control programs that manage a computer's operations and determine which other programs can be run. Operating systems thus define a computer's practical capability. They have also historically been, and largely remain, proprietary, with the result that application vendors must develop as many versions of their software as there are different machines they want it to run on. Nor can we expect all the programs that run under a vendor's existing operating system to run under a new operating system introduced by the same vendor.

At the application level, data base management systems vary in the range of operating systems they can run under and in how they represent information at the machine storage level, while different word processing systems have their own conventions for identifying type style and font, end of paragraph, footnotes, and so forth. Telecommunications adds further sources of incompatibility. Local area networks that provide departmental communications may not be able to link to one another, even within the same building. Token Ring, Ethernet, ARCNET, and other LANs are driven by different software management systems, and wide area networks that communicate across locations, firms, and countries have widely differing protocols and message formats.

Problems of incompatibility are often compounded by two common and essentially political aspects of IT in large organizations.

First, any effort to impose IT choices on decentralized business units can obviously become an intensely political issue. Each extension of reach or range relies on affected departments adopting common standards though not necessarily common equipment and software. It may also require modifying or even replacing incompatible systems. Departments may view such efforts as intrusions on their autonomy, particularly if they have a large investment in the existing facilities.

Consider the firm that has installed Digital Equipment Corporation computers for its engineering functions, Hewlett-Packard equipment for manufacturing, and comparable but incompatible local area networks of Apple, IBM, and Wang personal computers for various support functions, and has adopted the international telecommunications X.25 standard to link its offices worldwide. Just a few years ago, there was little need for separate departments to interlink their systems. But with IT becoming increasingly pervasive in communication, coordination, and customer service, pressure to integrate some, many, and sometimes all IT resources is mounting. For many firms, technical fixes may be practical but expensive. And who is going to tell these units that they must conform to a set of new corporate standards, which for many would mean essentially starting over?

The reality today is one of multiple, incompatible systems, growing operational needs to rationalize them, and recognition that integration is essential in the longer term. Senior management's responsibility is to define the degree of reach and range its firm needs to support the development of new applications and the cross-linking of existing applications and locations, and determine how essential it is to force the pace of integration today, by insisting on the creation of a shared platform based on what is practical now and incorporating standards that facilitate more open systems. These determinations should be based on business, not technical, considerations. Lacking senior management policy, progress is likely to follow the path of least resistance—the IT planning process will continue to be driven by local "now" issues and integration will remain a lower priority than meeting business units' urgent needs for new systems. Listen to a few of the IS managers who are caught in this situation.

> I am sitting on a time bomb. [The business units] are racing ahead to develop new dealer systems and new engineering systems and pushing the state of the art on computer-integrated manufacturing. The CFO is pushing worldwide finance and reporting systems. The pieces do not fit together at all. Yet, already finance is raising the issue of linking credit services to the dealer network and CIM must eventually backward integrate into engineering. There is no overall master plan and no momentum for one. I have not found any hook for getting the various parties to consider this as a priority.

> We cannot have a real office systems strategy until we resolve the issue of vendor X versus vendor Z. . . . There are armed camps out in the businesses ready to fight this war.

> Senior management appointed a special new group to look at European technical computing needs. The decision is a big one and it will be made in a relative vacuum. Any discussion of corporate "architectures" is seen as obstruction. I have no cards to play.

> The research division has become a set of [vendor X] shops, which is entirely incompatible with every other division. This prevents us from creating a common information and communications base. We know what is needed to rationalize the situation, but we can't wrestle things away from the scientific users. If information technology in general is unimportant on the corporate agenda, telecommunications is the least important. They are both high on the business-unit agenda and that's where the decisions will be made. . . . Yes, it will remain vendor X shops there and IBM shops elsewhere and any shop anyone else wants.

"So what?" is the reply of many business managers, who view the IS managers' complaints as turf-related. From a top management perspective, delegating decisions about IT to the business units is appropriate, reduces stress (on top management), and saves time and debate.

Reach and Range as Business Enablers and Disablers

There is no pat answer to the "so what?" question, but basic and consistent trends in business strongly suggest that firms will gain substantially, in terms of cost and competitive strengths, by making integration a new priority now: the longer they wait, the more the likely loss. Every large firm in every industry has seen how competitive and operational necessity has made it essential to put core transaction systems on line. The same logic leads them to move more processes on line. The bank that has introduced ATMs for customer service will add on-line balance inquiry systems, mortgage payments, and so on. Each of these may be handled by separate business departments so that there is no problem created by their being built on incompatible IT bases. At some point, however, it is almost certain that some of these core business services will need to link together. The move toward just-in-time everything, cross-functional coordination, and customer and supplier cooperation guarantees that. At this stage, the firm will wish the need for integration had been anticipated earlier.

The concept of an IT platform is based on the argument that the above sequence, by which more and more separate processes move on line and many of them then need to be combined, is inevitable. Many IT managers share this view. They express it in terms of the importance of "business process re-engineering" or of "integrating our business processes." They often jump ahead and talk of integration as a goal without helping business executives understand the business sequence that makes integration a technical necessity. However, it is not integration that is in itself the priority so much as making sure that likely business needs that depend on technical integration are enabled, not disabled, by having a platform in place instead of a set of separate IT bases.

Reach and range of the IT platform will significantly affect the degrees of freedom a firm enjoys in its business plans. Consider:

- the bank precluded from making electronic funds transfer at point of sale (EFTPOS) part of its business strategy because of the separate bases that do not provide efficient on-line connections to retailers' systems;

- the manufacturer locked out of opportunities to create joint efficiencies by separate bases that, although effective in meeting internal needs, cannot link to suppliers and customers;
- the automotive firm that limits both its organizational choices and the range of electronic partners with which it can form alliances for electronic data interchange, distribution, and computer-integrated design and manufacturing by not developing an international telecommunications and processing infrastructure; and
- the insurance company that cannot make the shift from a product-by-product strategy to a relationship-based strategy built on cross-selling and integrating information about customers, their demographics and use of the company's services, and the profitability of the relationship, because it lacks the necessary range in its IT platform.

The more information technology becomes a crucial element in firms' competitive and geographic positioning, organizational redesign, and human capital redeployment, the more dependent they will become on the characteristics of their IT platforms, and the more instances we will see of firms constrained from pursuing their long-term strategies by inadequate IT resources.

Connecting versus Communicating

For some firms, telex and fax may be adequate for communicating with other organizations. But these media will not allow such firms to link their purchasing and accounting transaction processing systems to those of other firms. This capability must be built into the design of the IT platform. As a business option, it may be a competitive opportunity today, but it is likely to be a competitive necessity by 1993 or 1995. Such capability is an extension of reach. It may also be desirable to add range, for example, providing for customer order information received by the sales system to be used automatically by the financial and manufacturing systems.

Extending reach does not compensate for lack of range or vice versa. Nevertheless, problems of reach are being alleviated by technical developments. In terms of equipment linkages we are approaching the same level of connectivity we now have in international telephones: e.g., direct dial to Japan, Britain, Egypt, and the Soviet Union. But direct dial is not the same as direct communication. I can phone Herr Gunther Beckard's office in Frankfurt and ask, "Is Gunther in?" But if the secretary does not speak English, I cannot exploit the connectivity unless I can speak German or find a shared language. Extension of reach does not solve the IT equivalent of talking to a German when you only speak English. That demands additional range, which is often impractical to provide or even infeasible, just as today there is no automatic English-to-German translation system.

Senior managers are likely to be puzzled by this situation. They may blame their telecommunications and information systems staff, especially when they hear from vendors, consultants, and academics that "open" systems are practical and guarantee extended reach and range. Open systems are analogous to today's photocopiers; they are very different in terms of design, features, and internal structure, but any standard-sized document can be copied on any of them. All the standards needed to guarantee extensive reach and range have been defined.

Defining standards is not the same as implementing them. The VHS/Betamax confrontation serves as a useful warning. VHS and Beta are standards. Beta had the support of a giant, Sony, but lost out in the marketplace. VHS won but, though standardized internationally, is incompatible across national TV systems. The British television standard, for example, has 50 percent more lines on the screen than U.S. sets; thus, a British VHS tape will play on a U.S. VCR linked to a U.S. television set, but all you will see is a jumble. Magnify this example many times, and you have a hint of the complexity of problems of reach and range in a large firm.

"Open" versus "Proprietary" Standards

Standards are key to integration. Advantages of features, cost, scale, and functionality can be played off in a vast range of hardware, software, and local and wide area telecommunications choices, just as the VHS standard has created a mass market for VCRs and tapes. Positioning the IT platform requires narrowing the set of choices and restricting local options to exploit existing and emerging standards. This can easily look like technocratic bureaucracy. Positioning also involves selecting one vendor or a few as the strategic suppliers and eliminating others. This can seem like compounding the bureaucracy, especially since it invariably results in an IBM/not IBM debate centered around open versus proprietary systems.[2]

Open-system standards are generally contrasted with proprietary standards that apply to a single vendor's products. An everyday equivalent is ribbons for impact computer printers. All perform the same function—transferring ink to paper—but most are specific to a particular printer. A ribbon for one printer will not work in another. By contrast, photocopier paper works on any standard copying machine.

The topic of open and proprietary systems is central to the evolution of the IT industry and impossible to discuss without discussing vendors. Many technical experts contrast open systems with IBM's proprietary systems, yet IBM's standards are often more open than open standards. An open standard is one that is fully defined, stable, and published in a form that allow vendors to incorporate it into their products. Many of the newer standards needed for open systems are not yet completely defined or stable, and an individual vendor's ability to implement them varies—rather like having photocopying paper that is "almost" 8 $\frac{1}{2}$" x 11". Many of IBM's competitors have built their products to fit IBM's main architectures, particularly its Systems Network Architecture (SNA), the cornerstone for achieving reach and range in IBM-based systems. SNA is in effect open in implementation. The MS.DOS standard, which has been the cornerstone for the growth of the personal computer industry, also began as a proprietary standard that is now effectively open. The many IBM clones that offer "IBM compatibility" suggest that the practical issue for managers is less standards in definition than standards in practice.[3]

IBM's dominant market position made it indifferent to standards other than its own. Hence it was a late arrival to the movement for open systems. It was particularly late in areas relevant to reach. The SNA architecture provided reach across the firm's IBM-based or IBM-compatible facilities, but not, for instance, across international customers and suppliers. Other standards provided this extra reach, though generally with very limited range. The X.25 standard for telecommunications, for example, has been widely adopted by value-added networks, national public data networks, and many computer vendors because it provides a fast and easy way to add reach for certain types of applications.[4]

Yet, IBM planners failed to realize until the late 1980s how important X.25 would be for many of its large customers. The corporate computing facilities of the 1970s did not need external reach, to other firms or international locations, for processing transactions. IBM also overlooked the growing need in the mid-1980s for fast and simple connectivity for departmental systems. Many of its more recent product announcements now aim at extending the reach of its main products, adding X.25 links, connecting to the local area networks customers are likely to have in place, and supporting major telecommunications standards. In 1990, it introduced several hundred products for "multivendor" integration; these conform to OSI, TCP/IP, and other widely adopted standards.

Many of IBM's competitors, while outpacing it in terms of reach, largely ignored the need for range. They provided easy connectivity for simple, low-volume communications, but failed to accommodate the more complex tasks of managing large volumes of information and handling high-volume transaction processing. While they exploited many open communications standards, their operating systems—the major determinants of range—were generally proprietary.

Neither SNA nor any comparable vendor architecture provides reach to "anyone, anywhere," that is, range across all information and services. This is the goal for Open Systems Interconnection (OSI), a comprehensive set of standards that has evolved through the 1980s and will continue to evolve probably well beyond 2000 before its goal is achieved. The world is moving very slowly toward open systems. Consequently, defining an architecture for an IT platform and choosing standards remain closely linked to choice of vendors. Because IBM standards form the practical base for the IT platform, all non-IBM providers must ensure that their products can interlink with IBM's. For its part, IBM must ensure that its products can interlink with those of leading competitors and that it moves at the same pace as the rest of the industry in implementing OSI. Market forces, user needs, and technical innovation converge. The practical path to open systems is via IBM architectures. This does not necessarily mean IBM products, any more than adopting MS. DOS as the standard for personal computers means adopting IBM machines.

Platform Costs and Cost Savings

The CEO who is less concerned about IT costs than IT value and fails to view IT as an important element in the firm's competitive positioning is likely to encourage full freedom of local choice. This is in effect saying: "Choose the technical option that meets your business needs now in specific areas or that gives us real cost-savings across operations now."

In practice, this seemingly reasonable mandate does not create the business flexibility, responsiveness, and sources of innovation ensured by the platform approach. Nor is it clear that it saves money in the long term. Pressures for the case-by-case approach often reflect short-term cost concerns. By definition, a stand-alone, special-purpose IT solution will almost always be less expensive and more efficient than one that is adaptable to a complex variety of requirements that must be cross-linked if not fully integrated.

Although the costs of building the requisite infrastructure are often initially high, the IT platform frequently reduces operating costs in such areas as telecommunications, shared resources, and economies of scale and expertise. Today many companies are spending large sums to hire systems integrators to link incompatible systems. It would have been less expensive to have built them within an architectural framework in the first place.[5]

Senior Management Policy Criteria

The path of least resistance in choosing an IT strategy is bottom-up, application-centered planning. In terms of future corporate business needs, it is not hard to make the case for such an approach. Senior executives neither can nor should be involved in the technical details of the firm's IT strategy. A set of policy-level requirements can be identified that provides a base for assessing a firm's existing IT capabilities and deciding whether there is a compelling business need for an IT platform. These policies can also help a firm's IT planners form a clear business framework and political mandate for designing a technical architecture for the platform built on a set of standards that imply criteria for selecting strategic vendors. The decision-making sequence for positioning the platform thus moves from business vision to policy to architecture to implementation.

The policy criteria that follow are essential to avoid being put at a competitive disadvantage. They incorporate the reasons an informed CEO is likely to insist on a platform that pushes as far along the reach/range combination as practical, while balancing technical risk with business risk and cost. No vendor has yet achieved either total integration or total openness. Hence there is inevitably some risk in choosing one vendor or standard as the strategic base for the IT platform. The business risk is the cost of loss of freedom, especially not being able to move with or ahead of the competition in areas in which IT is a necessary driver or supporter.

Policy 1: Business Practicality

Our IT base must never block a practical and important business initiative.

Most firms are paying more attention to the link between business needs and IT capacity and capability than they did when computers were confined to clerical operations and telecommunications meant telephones. In many instances, the IT links needed to enable or support a business initiative can be obtained from elsewhere. For example, a company that wants to establish links to its main suppliers for data interchange can use a value-added network service (VAN) supplied by companies such as GEIS or INFONET, or develop its own software but rent telecommunications facilities from, say, TELENET or MCI.

So long as a practical and important idea can be built on its own independent IT base, the specific choice of technology is irrelevant to the wider context of the firm as a whole. However, the inefficiencies and operational complexities of maintaining, say, four to forty different technical systems, with different skill needs, operating requirements, equipment, and software is in itself a problem; when they need to be made interdependent, the existing separate IT bases can be a blockage to business development.

Many likely business initiatives now and throughout the decade will involve linking firms' previously separate IT resources, such as manufacturing and financial systems, with their own and other firms' payments, materials management, stock forecasting, and EDI services. This requires additional reach. At the same time, multivendor systems must be able to interlink and connect with the systems of other firms, and fundamentally different technologies, such as image-processing systems and data base management systems, must be able to share information. This requires additional range.

Today's systems cannot do this automatically. Better IT planners will have anticipated these needs and made them major considerations in their choice of vendors and of standards. In doing so, they will have tried to balance short-term operations and cost pressures with provision for longer-term integration. Vendors, systems, and standards best suited to range are almost certain to limit reach and vice versa. Given the context, architecture planning today is at best an art form that involves complex trade-offs and judgments.

The problem is not as much reach as range. Reach can almost always be provided, if not always efficiently, through established telecommunications standards and gateways. Range is almost impossible to ensure without a firm commitment to a platform, and even then is a major challenge. The details of information sharing across technology bases are extremely complex. They relate to the basic design of the underlying computer systems, operating systems software, codes, technical data structures, message formats, and so forth. It is currently impossible to guarantee that an application built on vendor X's technology base can share information directly with an application on vendor Y's. Indeed, it often happens that such information cannot even be shared with a vendor X system that uses a different component.

There is another approach: avoid any business initiatives that involve the interdependencies that create the problem. This makes the technology base simpler to build and operate and almost certainly less expensive, and allows the CEO to leave major IT decisions to the business units. But this is a sensible option only if EDI, POS, customer-supplier linkages, computer-integrated manufacturing, and cross-linking of individual products and services are not likely to be in the mainstream of the firm's activities within the next five years. It is hard to find an industry in which this is likely to be the case.

Policy 2: Competitive Lockout

If our competition uses IT as the base for a successful initiative, we must not be automatically locked out of countering or imitating it.

Leading firms in many industries, even firms with strong IT capability, have been unable to respond to IT-based initiatives by competitors during the past decade. Delta, Johnson & Johnson, and Bank of America, to name just a few, were left at a competitive disadvantage by American Airlines, American Hospital Supply, and Citibank, respectively. The ability to respond to the moves of competitors depends both on having IT skills in place and on the technical bases of the competitor and the company looking to respond. An initiative based on an off-the-shelf software package can be countered easily by buying the same package. A space-age application that many other firms in the industry can copy is less likely to yield sustained competitive advantage than a platform-dependent application that few or none can match.

There is an immense difference between a competitive IT application and a competitive platform. A major weakness of the extensive academic and popular literature on the links between IT and competitive advantage is that it ignores the actual technology. Technology must be put back into discussions of IT and competitive advantage. Senior managers are aware that information technology is at least potentially linked to competitive positioning, that it is no longer an overhead function. That awareness opens up new possibilities for action. Without it, managers are blind to both possibilities and necessities.

Ironically, raising senior managers' awareness about IT can decrease their understanding of the importance of information *technology*. The main reasons for this are that:

- almost none of the cautionary tales and examples describe the technology base used to obtain the competitive advantage, implying that the choice of technology is irrelevant;[6]
- few identify precisely what establishes the advantage, hiding the difference between a competitive application and a competitive platform and between an advantage and a sustainable advantage; and
- awareness is passive, removing roadblocks to changing how IT is used but not providing a guiding vision for IT planning.

If we ask, "What difference does the choice of technology make?" many of the old stories take on new meaning. Merrill Lynch's Cash Management Account, for example, relied on linking its systems and those of Banc One of Ohio. The cornerstone of CMA was a Banc One Visa card and a transaction processing system that included a "zero balance" account that cleared all CMA transactions daily. CMA could have been a major fiasco had Banc One not been able to adjust smoothly and quickly to the totally unexpected level of demand. Merrill Lynch had forecast about 40,000 customers for CMA. At one point, it was signing up 20,000 accounts per month. Eventually more than one million accounts were established without exhausting capacity.

An example of a comparable business initiative by another frequently cited exemplar, Citibank, shows how a convenient, cost-effective, short-term technical choice can be a long-term mistake. The special-purpose hardware Citibank used to implement its international COSMOS system permitted rapid development of small-scale applications and low-cost operation but could not handle large volumes, necessitating a change of vendor at great expense and loss of momentum. Additionally, Citibank had to run two versions of COSMOS, one under each of the new vendor's different operating systems for mid-sized and large computers.

The AAdvantage frequent flyer program is widely cited as an IT innovation that changed an industry. Sabre, AA's reservation system, and United Airlines' Apollo are both viewed as making IT a central element in airline marketing and distribution to a degree that computerized reservation systems (CRSs) have become the central battleground domestically and internationally.

AAdvantage succeeded mainly because it was based on the Sabre platform, which enabled American to cross-link passenger data from Sabre to AAdvantage to its yield control systems and to its hub planning systems. Its competitors could not do that. They had the same applications as American but not the platform that integrated them and that allowed—and still allows—American to add application after application, with cross-application linkages an automatic by-product. A manager in one of American's leading competitors describes its equivalent frequent flyer program—a separate application with the same degree of reach as its CRS but without the range to cross-link with it—as an "albatross," all cost and no benefit. Sabre not only created a distinct competitive advantage for American; it also imposed a competitive lockout on most of the airline industry.

Policy 3: Electronic Alliances

We will match the competition in being able to make alliances, create value-added partnerships, or enter consortia.

Increasingly, nontechnical standards for business interchange are becoming as important as the technical standards and architectures upon which IT platforms are built. Many of these industry-specific standards, which define formats for transaction documents and procedures for electronic processing, are defined by industry associations. They include the U.S. grocery industry's uniform communication standard (UCS), the international banking community's SWIFT standard, and comparable standards in transportation, warehousing, automobiles, and aerospace. The international EDIFA and closely related U.S. X12 standards are the basis for many of these business standards. It is no exaggeration to say that EDIFACT is as important as SWIFT in creating a base for entirely new electronic markets and "metabusinesses."

These interchange standards are critical requirements for inter-company uses of IT such as point of sale, EDI, customer/supplier links, and the creation of new cross-industry services such as hotel and car rental bookings provided with airline reservations. As more businesses move their core operations on line and create electronic links to other firms, they will choose electronic partners and be chosen by them on the basis of IT reach and range and interchange standards. One of the largest U.S. banks, for example, lost a $400 million client because it could not link its payment systems to the customer's as part of an electronic supply system that used General Motors' Manufacturing Automation Protocol (MAP). A firm whose IT base meets only its internal needs will be at an obvious disadvantage.

In the 1990s, on-line operations will naturally, rapidly, and inevitably be extended across organizational boundaries, with IT as the enabler. This poses very new challenges for both IT managers and business executives. What range of business interchange capabilities must they plan for? Must a bank's platform be designed to ensure transactions and information sharing with petrochemical companies and auto makers for point-of-sale and electronic payments? Should an insurance company prepare for electronic data interchange for all employee benefits paperwork and payments involved in servicing corporate clients? Should a parts supplier assume that its engineering department must be able to interchange technical specifications with vendors and customers? Should a bank adopt the MAP standard that General Motors requires its suppliers to use on the premise that banks (and GM's insurance providers) will have to provide the same electronic linkages? Since there are no widely established standards for graphic design documents, which ones should a manufacturer monitor and be ready to adopt at the appropriate time? Should pharmaceutical firms agree on common standards for image and electronic data interchange so that they can propose them to the FDA and speed up the complex certification process for new drugs?

The answer to all these questions is surely "yes." They do not necessarily relate to specific applications; they are quite literally a platform on which to build applications. They are part of long-range corporate planning.

Policy 4: Reorganization and Acquisitions

If we reorganize, make acquisitions or divestments, or relocate operations, our IT systems will adapt quickly and routinely.

The first three policy requirements for the IT platform relate to ensuring firms' competitive health. This policy relates to firms' organizational health. Every CEO of a large firm assumes that mergers, acquisitions, and divestments are likely to occur. Reorganization, relocation, and related organizational changes are even more likely. Ideally, the IT platform should provide an electronic organization structure that can shift as the organization shifts. A company that acquires another firm is very likely to have IT applications that differ significantly in terms of hardware, software, and (most constraining on integration) information formats and structures. The ease with which the acquiring firm can mesh the acquired firm's systems into its own technology base may be a major determinant in its decision whether to make a deal.

Two U.K. financial service companies, for example, agreed to merge and then cancelled the decision by mutual agreement when they found that their IT systems could not be combined and interlinked. Instead of creating a company with advantages of scale and joint operations, they had two separate companies within a company. Their operations and information systems were so tightly interdependent that without a common platform for linking the IT base they could not link the operations base.

The almost certain future importance of IT in acquisitions decisions strongly suggests that firms keep their IT architectures as close to the mainstream as possible. Yet very few IS and business managers have even a clue about customers' and competitors' architectures.

Policy 5: Third-Party Intrusions

No firm in our industry, or third parties outside it, will be able to intrude on our areas of strength or into the mainstream of our marketplace.

One of the distinctive features of IT is the degree to which it erodes and even eliminates boundaries between industries. The most obvious example is electronic funds transfer at point of sale

(EFTPOS). Is this retailing or banking? Publix, a Florida supermarket, used its own POS base to become a leading supplier of banking services at the expense of the major state banks; it franchised its platform out to smaller banks and savings and loans and other financial institutions.

Airline reservation systems similarly now include nonairline services, and customer/supplier distribution and EDI systems incorporate electronic payments. A firm that has a first-rate platform with extensive reach will naturally look for opportunities to add traffic. If that platform combines extensive reach and range, the firm may be more effective in some areas outside its industry than firms in those areas. For example, international airlines' CRS capabilities are outstanding, but their systems for handling payments for bookings between airlines and travel agents are primitive. Banks in several countries have proposed replacing the airline bank settlement plan with their own electronic payments systems. This would not entail new facilities, only an extension of the banks' processing and communications base to the airlines.

McKesson was an early exploiter of this opportunity. Its initial innovation was to link its customers, small pharmacies, to its purchasing, distribution, and inventory management systems. It has since added insurance claims processing. McKesson had the reach into the pharmacies that the insurance providers did not and extended its range to add claims processing as a by-product to its mainstream transactions. It is now the third largest claims processor in the United States. All that business once belonged to insurance companies.

Firms must assume that, just as their partners may come from outside their industry, so too may their competitors. The IT platform must ensure defensive capability against third parties.

Policy 6: Vendor Staying Power

We will not be dependent on "brochureware" nor on vendors that will not be able to stay the course.

"Brochureware" refers to the claim that "we have it all: OSI, integration—you name it." Many previously strong IT vendors lost ground badly during the past decade because they either failed to provide extensive reach and range in their products or lacked the R&D base and funds to move with the mainstream of the industry toward integration and implementation of key standards. Wang,

Data General, and Prime are examples of vendors that have lost their position as strategic suppliers for these reasons.

Early 1980s leaders in data base management systems have similarly been unable to broaden their products to accommodate large firms' current needs for reach and range. Software providers in the personal computer field have experienced increasing problems and delays in upgrading their products to handle a plethora of telecommunications and multivendor systems. Cullinet's fall from leadership to acquisition and dismemberment is an early signal that product features are not enough; proven ability to move with the mainstream of standards and integration is now essential. This is beyond the capabilities of all but a few vendors, but it is in the brochureware of many. Selecting standards that will help a firm evolve a truly integrated platform means selecting vendors that can stay the course.

Policy 7: Comparable International Capability

The preceding policy requirements will be applicable in an international context.

It is vital that firms with an international strategy for competing in a global marketplace have an international IT strategy and an IT platform that extends worldwide. Very few firms have such a strategy, and very few vendors can fully support it yet. International telecommunications is a political, regulatory, and technical minefield. In the 1990s no major firm will be able to operate efficiently in a global environment unless it has a global IT platform that ensures international reach, which means adopting international telecommunications standards, and sufficient range to support information-sharing across all functions and locations.

Reuters stands out in having created a worldwide market through its platform. The *Economist* summarizes the firm's accomplishments.

> Reuters has effectively become the global market for currencies, carrying price-bending news as well as banks' quotations. . . . [After 1981] some Reuters terminals allowed dealers to execute transactions with other subscribers—a kind of souped-up telex.

> The new system, named "Dealing 2000" . . . will allow traders around the world to enter orders into the system, which will then match them automatically.[7]

Multinationals will increasingly need to create their own equivalents of Reuters' global information and communications delivery base.

Issues for the IT Planner in Defining the Platform

The technical base for the IT platform is defined in terms of an architecture built on a set of standards. The architecture must balance a variety of needs, all of which cannot, in reality, be fully met. It must:

- provide the maximum practical degree of protection of existing IT investments;
- ensure that the firm maintains the opportunity to adopt new technology;
- define a longer-term path toward integration of all relevant components of computing, telecommunications, and information management resources; and
- accommodate emerging open-system standards and emerging technical and business standards that extend the firm's platform to other companies.

Applications designed around an architecture and based on its technical standards will be able to share information automatically and directly with other applications being run in other locations, consistent with the degree of reach and range built into the architecture.

Standards: The Feasible and the Practical

The goal of the corporate IT platform is to share resources to gain economies of scale, combine information from separate applications in order to create new services and products, avoid redundant and duplicate facilities, use the same delivery base for a growing range of services, facilitate cross-organizational information flows, link the firm's systems with those of customers or suppliers, or some combination of these. These objectives assume a high degree of compatibility among computers, software, information resources, and telecommunications, and integration of separate components of the firm's existing and future IT resources. Establishing a single and monolithic facility is not required for this; ensuring that the technical components share common standards is.

Consider the everyday example of the telephone system. When you dial a number, electrical signals are sent to the telephone exchange to establish a connection. The receiving device recognizes and interprets the signals and responds. The sending and receiving systems can communicate because they share the same standards, even though they use entirely different voltages, signaling equipment, and so forth. An even more mundane example of a standard interface is the two/three-pin plug we take for granted when we plug an electrical appliance into the wall. If only the computer and telecommunications field had evolved along the same path as telephones and appliances!

The new imperative in the IT field is to end incompatibility by moving as quickly as possible toward industrywide standards and integration. Standards are key to reducing incompatibility. The rise in popularity of UNIX as an operating system rests on the claim that it is fully portable across hardware environments; a program developed on equipment from vendor *A* can be "ported" directly to vendor *B*'s equipment. (In practice, this capability is limited today by variations in UNIX implementations.) Definitions of the components of the open systems that will provide "anyone, anywhere" integration are still mainly in the proposal stage. It may take several decades to realize this capability, although progress is accelerating as business demands an end to incompatibility and vendors accept the fact that their systems must be able to interlink with those of other manufacturers. Some of the main areas in which standards are needed to reduce problems of incompatibility are summarized in Figure 7-2.

Being developed in parallel with open systems is the Integrated Services Digital Network (ISDN), the worldwide and fairly coordinated movement by telecommunications providers to bring together on a single transmission facility every type of information: telephone calls, documents, pictures, video, and computer data. Just becoming available in the United States, France has operated ISDN since 1987. But already the various pilot versions of ISDN are mutually incompatible, and the technology has moved so much faster than the standards implementation process that ISDN is already obsolescent.

Standard-setting organizations can work five to ten years just to reach agreements. Vendors then have to embody the standards in their products. In the meantime, firms must decide which emerging standards to bet on and how to deal with existing proprietary systems.

FIGURE 7-2 **Needs for Standards to Ensure Integration**

	Competing Propri- etary Systems	Base for Move to Open Systems
Operating Systems	MVS (IBM) VMS (DEC) plus many others	UNIX
Personal Computers/ Workstations	MS.DOS Macintosh	UNIX
Local Area Networks	Ethernet Token Ring	TCP/IP
Telecommunications Architecture	SNA	OSI SNA
Data Base Management Systems	Oracle DB2	SQL
Electronic Mail, Telex	Many	X.400

Choosing Standards and Evolving the Architecture

Choices of standards define the architecture for the IT platform. A firm's architecture is, in effect, its technical strategy. The word "architecture," which has become a central part of the IT planner's vocabulary, is conceptually vague and has many subdefinitions. An architecture is a blueprint rather than a facility. It is often compared to the city plan that lays out major highways, sets zoning ordinances, and defines locations and utilities. It does not describe the details of houses, though it may impose standards of size, construction, and safety.

The functionality of a firm's IT platform is determined by its combination of reach and range. These are determined, in turn, by the standards on which its technical architecture is based. Standards depend on vendors' products and systems. Though choosing a standard is not the same as choosing a vendor, until open systems are proven and comprehensive, the choice of standards and strategic vendor are highly interdependent.

Views on vendors and open standards are widely divergent, and much of the discussion centers around IBM. Roughly, there are two extremes of viewpoint.

Many telecommunications professionals, computer scientists, and technical specialists in engineering and manufacturing, whom we'll dub "open systems purists," argue that OSI, UNIX, and other more recent standards are essential vehicles for creating a platform capable of more fully exploiting IT. They view IBM as a laggard that has been historically concerned primarily with protecting its markets through proprietary architectures, and they point for evidence to the complexity of many IBM products and architectures that are based on 1970s technology. They tend to focus on reach as the key issue for integration, and they promote standards that provide mainly connectivity. The vendors they regard as strongest are firmly committed to open systems and able to provide easy connectivity.

The "corporate realists," a large plurality of IS managers, view IBM's architectures as the essential base for a corporate IT platform. They see OSI as still a long way off, while SNA is already here. And they recognize that IBM is the de facto standard in personal computers, telecommunications, and data base management systems. Their priority is range—integration of information rather than just extension of connectivity—and they see IBM as far stronger in this area than its competitors, few of which can provide both operating systems and applications software capable of handling the massive data bases and large volumes of transactions that are characteristic of business environments.

The corporate realists accept the fact that IBM has ignored many important aspects of connectivity—only recently, for example, has IBM shown real support for X.25, OSI, and Ethernet—and they are impatient with IBM's slowness in responding to these needs. But they believe that adopting IBM architectures and including components from non-IBM vendors afford the widest spectrum of choices. IBM's 1990 introduction of its powerful UNIX-based workstations and OSI products demonstrates that IBM now firmly accepts that any leading IT provider must accommodate multivendor reach and range in its products.

Senior managers may be puzzled by the intensity of feelings among IT managers, academics, technical specialists, and consultants in the ongoing and often acerbic debate between purists and realists. What they are seeing is a reflection of the importance of the issues, complexity of the topic, and commitment to achieving the ideals of integration and impact on business and society. Howard Frank, one of the most experienced consultants in the IT field, observes that IT planners are now having to solve problems of disintegration created in the 1980s by short-term, reactive IT planning, problems that were widely foreseen; leading IS managers and practitioners have been talking for years about the need for architecture. They were unable to convince business executives of the urgency of the problem.

Positioning a firm's IT platform must address today's realities as well as tomorrow's possibilities. The reality in most firms is a multivendor, multitechnology, hence multistandard IT base that must be integrated as quickly as practical without discarding the pieces and starting again. Rationalizing current systems is a first requirement.

The first step toward integration is to adopt a target architecture based on an overall set of standards that are fully implemented today. These are almost certain to be at least partly proprietary. Evolution toward open standards can occur as they become more fully defined. The need to balance today's operational needs with longer-term movement toward open systems and integration means that the criteria for selecting strategic vendors must go well beyond which ones have the "best" technology. Also to be considered are:

- the ability to help rationalize today's installed IT resources;
- proven support for today's mainstream proprietary and nonproprietary systems;
- proven capabilities in implementing new standards;
- breadth of capability in all areas relevant to the IT platform, including telecommunications, computing, and information management; and
- the ability to upgrade facilities smoothly and quickly and to perform efficiently as processing volumes and transaction traffic increase.

Three critical elements are largely ignored in most discussions of standards, integration, and open systems.

1) Open and proprietary definitions of standards are different from open and proprietary implementations of standards.
2) The costs required to turn the definition of a standard into software and/or hardware are the same for a vendor with a base of 10,000 customers as for one with a base of 10 million. The software development costs involved in moving along the path to integration and the promised land of open systems are far greater than most vendors will be able to bear.
3) The process of implementing standards has become clearer in recent years. To reach a critical mass in the marketplace and

be considered a primary supplier of most large organizations' information technology base, a vendor must be able to offer IBM compatibility and/or "interoperability"—the ability for the different systems to work together reasonably efficiently.

IBM has made it clear that it will no longer risk being locked out of any major emerging market; it is moving toward interoperability with all relevant developments. This means that the path to open systems in practice is via IBM architectures. That does not necessarily mean IBM products; a firm can implement its IBM architecture by adopting, say, Amdahl host computers, Northern Telecom communications switches, and Toshiba personal computers.

Vendors who plan to go beyond brochureware have to be able to do the following over a decade or more of investment:

- support their existing product base;
- move with the pack in areas of open systems and related standards most immediately relevant to multivendor interoperability; and
- maintain compatibility with IBM's major systems.

Examples abound of firms that were strongly positioned in the early 1980s only to drop behind the leaders or out of sight altogether because they could not evolve their products along the "practical integration path," that is, accommodate emerging telecommunications, data management, and computing developments without requiring that installed capabilities be discarded.

Market forces have defined the practical integration path for vendors. It is driven by two main forces: open systems as an emerging direction for the entire IT industry, and IBM's architectures as a de facto standard. These two forces, in conflict until the late 1980s, are increasingly converging.

Apple's initial attempts to maintain the Macintosh as explicitly and deliberately IBM-incompatible failed badly. For several years, Apple was kept afloat by the venerable Apple II. When Apple finally accepted the vital need to provide linkage with IBM's mainstream products, sales of the Macintosh took off. Similarly, Digital Equipment Corporation in the mid-1980s switched from maintaining DECNET as a proprietary telecommunications architecture to providing SNA linkages. Approximately eighty vendors offer SNA compatibility today. SNA has become the gateway to, if not the information world, then the *Fortune* 1000's and its European equivalent's information technology base. This makes SNA effectively open. The same is true of MS.DOS; the openness of that proprietary standard gave rise to the IBM PC clones and an entire software industry.

The evidence suggests that, to get more than a small foothold in large organizations' IT resource base, vendors must shift to IBM for reasons of competitive necessity. The history of UNIX illustrates an alternative approach. UNIX is generally regarded as both relatively new and open. It is neither. UNIX was developed at Bell Labs in the 1970s. For many years, AT&T urged the adoption of UNIX and IBM ignored it. Only in the past few years have developments in microelectronics enabled UNIX to move out of the university hackers' world and onto the power platforms of the intelligent workstations. Startup firms such as Sun and Apollo raced ahead of established vendors in moving UNIX from the periphery of the IT field toward the center.

Noteworthy in the progress of UNIX in the United States is that:

1) the leading UNIX providers increasingly recognized intercon- nection with IBM's architectures as essential for reaching a critical mass in the corporate market; and
2) IBM, no longer able to ignore UNIX, spent $400 million in 1989 to create its own version, AIX, and early in 1990 launched AIX workstations.

The purists have largely ignored the message of this example—the inevitable convergence of open systems and IBM architectures, driven by technology and the marketplace. (See Figure 7-3.)

The term "SNOSI" ("SNA-into-OSI") captures the natural, pre- dictable, and practical path toward the multivendor environment that every customer wants and every vendor is aiming at. The prac- tical path to integration is now apparent: all leading vendors are moving roughly in parallel to arrive at open systems (OSI), UNIX, and IBM coexistence/compatibility. If this path is followed, we are likely to see the following.

- IBM's flagship architectures will remain the driving force in the industry (and firms such as Amdahl, NEC, Tandem, and NAS will continue to compete against IBM through those architec- tures).
- A new standard or architecture offered by any vendor intent on gaining a strong position in the corporate market will be

FIGURE 7-3 **Open Standards**

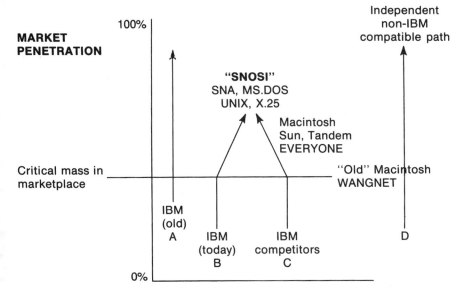

UNIX: Open Definition, Proprietary Implementation
SNA: Proprietary Definition, Open Implementation

adapted, by the vendor or a third party, to provide IBM compatibility.

- If any innovation reaches critical mass in the corporate marketplace, IBM will move to coexist.

The job of senior business managers is not to choose standards and vendors, but to sufficiently understand the issues involved in doing so to ensure that their IS managers see that decisions about standards and vendors are strategic choices, that choosing IBM architectures is not the same as choosing and being locked into IBM, and that there are no magic or easy answers to the problems of integration.

Enablers and Disablers: The Top Management Agenda

That firms' IT platforms will increasingly determine their degrees of freedom in the 1990s is clear. This makes the IT platform a major business resource. CEOs need to ensure that it is planned as such. Figure 7-4 provides a simple framework for mapping a firm's IT platform capability in relation to policy drivers and key business

FIGURE 7-4 Business Technology Platform Mapping

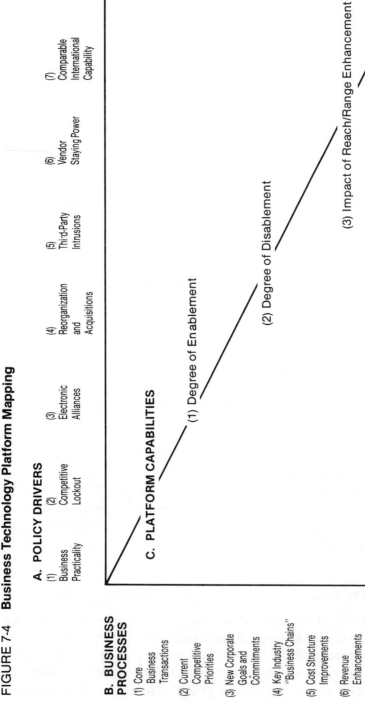

A. POLICY DRIVERS

(1) Business Practicality
(2) Competitive Lockout
(3) Electronic Alliances
(4) Reorganization and Acquisitions
(5) Third-Party Intrusions
(6) Vendor Staying Power
(7) Comparable International Capability

B. BUSINESS PROCESSES

(1) Core Business Transactions
(2) Current Competitive Priorities
(3) New Corporate Goals and Commitments
(4) Key Industry "Business Chains"
(5) Cost Structure Improvements
(6) Revenue Enhancements
(7) "Business Invention" Opportunities

C. PLATFORM CAPABILITIES

(1) Degree of Enablement
(2) Degree of Disablement
(3) Impact of Reach/Range Enhancement

processes. It can help a firm to answer questions relative to how well its IT platform is positioned in terms of:

- **Degree of enablement:** For each policy driver, and for each business process, what degrees of freedom do we have?
- **Degree of disablement:** Where do we run the risk of not being able to move ahead in our business, in terms of both competitive and organizational opportunities and necessities?
- **Impact of enhancing reach and/or range:** What impact would extending the functionality of the platform have on our degrees of freedom? (This answer relies on identifying where, when, and how using additional vendor products and architectures, or implementing key technical, industry, or business interchange standards, will add additional reach and range to the platform, and the impact these extensions will have on business processes and policy drivers.)

The business processes listed in Figure 7-4 are not mutually exclusive, and some key aspects of a business may appear in several rows.

1) *Core business transactions* constitute the basis of a firm's operations (efficiency, quality, and service). Failures here have direct and visible impacts on performance and reputation and, conversely, improvements afford long-term value.
2) *Current competitive priorities* are a firm's major stated priorities and critical success factors.
3) *New corporate goals and commitments* are the major business initiatives that are under consideration, planned, part of capital investment commitments, or more generally part of a firm's stated mid- to long-term strategy.
4) *Key industry "business chains"* are intercompany transactions that directly involve customers, suppliers, industry consortia, or third-party providers.
5) *Cost structure improvements* are major targets and plans for improving a firm's cost dynamics.
6) *Revenue enhancements* are major targets for increasing revenues through new markets, increasing market share, or increasing a firm's share of key customers' purchases.
7) *"Business invention" opportunities* are opportunities to launch radically new initiatives that are not part of current planning

and that do not fit into today's market segmentation, product strategies, and organization.

Mapping the firm's platform capability in relation to policy drivers and business processes is a necessary reality test for senior business managers. A manager who cannot answer the questions raised by Figure 7-4 needs to ask why. "Because I have not thought about IT in business policy terms? Because I do not understand how our key business processes relate to IT and IT to business? Because we do not have an IT platform?"

The issues discussed in this chapter have been a source of constant concern for leading IS executives, who recognize that the architecture is the technical strategy, that integration is essential, and that the lead times and complexities of IT planning demand clear and early business directives. That few senior executives are familiar with these issues may account for senior management's role as an impediment to ensuring that these needs are met.

Notes

1. Much of the material in this chapter addresses issues of technical standards. The whole field of standard-setting is extraordinarily complex, volatile, controversial, and above all packed with often bewildering technical terms. In order not to overload this chapter with definitions and jargon, relevant technical details have been described in the Glossary at the end of the book.

2. It is impossible to discuss IT standards without including IBM. Its dominance of the computing marketplace of the 1960s and 1970s means that most companies' basic accounting and related systems are likely to be built on IBM operating systems. That same dominance created "IBM-compatible" offerings from firms such as Amdahl, Hitachi, and NAS in the mainframe market, and from literally hundreds of companies in the personal computer market.

3. MS.DOS shows the power of the marketplace in deciding which standards will be adopted most widely. Kindly stated, it is user-hostile, arbitrary, overcomplex, intolerant of errors and novices, and badly designed. However, it became the standard across the personal computer field once IBM chose it over several now-forgotten competitors' operating systems. It created an industry. Even though IBM introduced an entirely new operating system on its PS/2 series machines, DOS is alive and thriving, particularly in the laptop PC market.

4. SNA and X.25 are the two key telecommunications standards. They are not at all equivalent. SNA is the overall telecommunications architecture for IBM's main

computing and communications offerings. X.25 handles only a subset of what SNA provides. The OSI model, which builds on X.25, is directly equivalent to SNA in functionality but only in terms of definition; many of its elements are either undefined or defined but not implemented.

In any discussion of telecommunications strategy in a large firm, particularly an international one, it will be unusual to go more than five minutes without hearing at least one of the three terms—SNA, X.25, and OSI.

5. In the late 1980s the conventional wisdom about IT tended to favor maximum decentralization—"distribution"—of technology and argue against consolidation and centralization. The obvious drivers here were the power and economics of personal computers and local area networks. More recently, the advantages of suitable centralization are becoming more clear. They include large data centers, which provide economies of technical expertise, shared telecommunications transmission facilities, and centralized network management systems. Instead of seeing centralization or decentralization as an either/or issue, IT planners now aim at both.

6. Hardly any of the literature on "competitive advantage" successes ever discusses any aspect of the technology. Many of the authors admit to being entirely unfamiliar with it. This has encouraged a trivialization of the immensely complex issues of technical strategy and design, even among heads of IS groups. Bruce Rogow, executive vice president at the Gartner Group, describes this as the "tinkerbell" syndrome. "The lemming approach causes a mentality of one-upmanship. . . . So it's like, I'm going to be more business-oriented than you are. Not only am I going to be more business-oriented, I'm going to ignore technology. . . . To survive the next decade [of massive technical shifts] CIOs will have to re-emphasize the technology. That isn't to say that they should ignore the business side; it's just to say that the pendulum has to swing back." (K. Melymuka, "Say It Ain't So. . . . OK, It Ain't So," *CIO* (March 1990), pp. 53–61.

7. Quoted in "Can We Talk?," *CIO* (May 1990), pp. 22–32.

Chapter 8

Aligning Business
and Technology

Managing change means reacting to it. It implies that change has a beginning, a middle, and an end, and that organizations can define an equally programmatic strategy for "unfreezing" the status quo, advancing the change process, and institutionalizing the changes that are made.

This chapter is not about managing change; it is about taking charge of change. Change cannot be "managed" when there is no longer a status quo, when waves of change follow one another. Under such circumstances, change sets its own pace. That change is the only constant is a cliché today, a cliché that seems more an expression of a sense of loss of control than a statement about "reality." Managers see change coming at them at a pace far faster than they can handle.

Of all the major business areas, only information technology seems to elicit from executives a passivity about their own response to change; many either half-apologize about being too old to adapt or half-boast about their lack of understanding of IT. IT is an area of continued frustration for them, for reasons of cost, elusive pay-off, disruption, complexity, and, perhaps most of all, the too often conflicting elements of change that attend it—too rapid technical change and too reactive and often mistimed organizational and managerial change.

Information technology has been dominated by the journalistic perspective, which makes change very much a "gee whiz" topic. Almost every innovation—expert systems, word processors, chips, supercomputers, videotex, smart cards—is viewed in terms of its future promise as part of a revolution.

A revolution has been going on, but in an evolutionary sort of way. Today's "symbol economy" dwarfs the physical economy. New York City's phones and data communications networks process \$1.5 trillion financial transactions a day. London processes over \$100 billion a day in foreign exchange. Sometime in the early 1990s, it will trade in a single day the annual GNP of the United Kingdom. At any moment, the air around us is literally full of microwave and satellite signals carrying the new currencies of foreign exchange, security sales, funds transfers, purchase orders, and the like. Little of this activity shows up in economic statistics, particularly GNP figures. Perhaps we need a new index, the Gross Information Product (GIP), to track the extent of the economic change IT has created.

Technology moves at a pace that makes it at times indistinguishable from magic. There is almost nothing one can predict about technology that is implausible. Late 1970s' plans for Integrated Services Digital Network (ISDN) set telecommunications transmission at 64,000 bits per second as a long-term goal and challenge. Fiber optics today supports speeds of well over one billion bits per second. The early computer chips that made possible the personal computer and its offspring had four thousand bits of memory. Today's temporary upper limit is four million bits. Sixteen million bit machines are already out of the lab and being readied for production. At the start of this decade, a 15 mips (millions of instructions per second) computer was a large mainframe. Today's high-powered workstations are rated at 40 mips. There is no reason whatever to assume this pace of technical innovation will decrease. Just as the innovations in the 1980s of personal computers and fiber optics were almost unimaginable by the practical business person, those of this decade are as yet unimagined and will astound us yet again.

Consistently over the past three decades, effective use of rapidly changing technology has lagged behind its availability. Frequently, technical change turns out to be easier than expected, organizational change significantly harder. Software development remains an art form. Getting business value from IT remains a promise of future benefit. It has been the lack of change that has caused problems.

Most of the management dilemmas and challenges discussed in this book are the same as were faced in 1980. Many are the same as in 1970. Figure 8-1 briefly summarizes these concerns. Three

FIGURE 8-1   **What Hasn't Changed: Management Challenges, Dilemmas, and Concerns through the 1970s and 1980s**

The items shown are taken from a range of surveys and articles over the past twenty years. They show management concerns that appear again and again and that do not seem to have been satisfactorily resolved.

	Whose Priority Concern Is This?	
	Business Managers	**Information Services Managers**
Setting business priorities for IT investment	X	X
Funding corporate IT infra-structures		X
Seeing real economic pay-off from IT	X	
Building relationships with top management		X
Educating top management		X
Making sure projects deliver on time and within budget		X
Controlling IT costs	X	
Aligning the IT strategy with the business strategy		X
Defining an architecture		X
Integrating separate technologies		X

circumstances have strongly contributed to managers' discomfort in dealing with IT as a force of "change":

- an uncomfortable relationship between business managers and IS managers and professionals;
- the lack of a coherent business model for mapping IT and its breakneck developments into that relationship; and
- a history of great expectations from IT, of grandiose promises and disappointing, sometimes disastrous, outcomes.

Key to getting unresolved management concerns out of managers' zones of discomfort is to replace old monologues by dialogue. In many respects, this entire book is about dialogues. Each chapter presents a management opportunity and challenge relevant to a particular aspect of business design through information technology—and a particular aspect of the business/technology dialogue.

If only the people who understand business, organization, and information technology thought alike, shared a common language, and talked and worked together! So many of the problems, disappointments, disruptions, and unmet expectations so common to IT result from misalignment between business and technology. IT successes generally reflect an effective relationship between business managers and Information Services managers and their staffs.

In the effective business/IS relationship, one party may lead and one may support. Business imagination may create an idea for innovation that technical practicality then turns into action. Alternatively, creativity on the part of the technical planners may be backed by business practicality. Rarely has one without the other produced results.

The key to alignment is relationships, not "strategy." There is nothing about IT that makes it any more difficult to manage than finance, marketing, production, or human resources. The real problem seems to be the history of relationships or lack of relationships in most organizations: the growth of the data processing and telecommunications professions as a technical elite isolated from the wider business; business managers' inexperience with and fear, suspicion, abdication, and delegation of IT; business units' dependence on a central IT monopoly and later rejection of it; and a mismatch between business and IT planning processes, accounting, responsibilities, and knowledge.

All these problems are now widely recognized and real efforts are being made to resolve them. Information Services units are being distributed out of corporate IS to business units. Steering committees and equivalent mechanisms are providing business inputs into what was before largely a technical planning process. Support of end-user computing has become a priority for IS. Many companies have initiated business management awareness education programs. And more and more heads of IS are coming from outside the field.

All of these actions help to foster dialogue and improve alignment, but the evidence suggests that, at the top, dialogue is still incomplete or even altogether absent. What there is at the top are two monologues, old monologues, that grow louder as the risks, levels of investment, and business impacts of IT grow.

There are many blockages to the needed business-IS dialogue, most of them products of the history of data processing and the resulting perceptions business managers and IT managers bring to their encounters. These old complaints are boring and their repetition does little to move the organization ahead. Yes, the IS organization has far too often not delivered on its promises. Yes, technical people too often do not understand business. Yes, users too often do not know what they really want, and senior managers won't make the tough decisions about long-term infrastructure needs. So what? Instead of dwelling on the old concerns, why not accept their historical validity and work on opening the needed dialogues that will end them?

Addressing the Issues

Part of the difficulty in communicating the meaning of the IS executive's concern about integration and infrastructures is that these topics are extremely hard to explain. The reach/range platform framework presented in Chapter 7 is one effort to explain integration in business terms and provide a base for linking business integration and technology integration and business policy and technology policy.

The business executive's concerns are less easily addressed. It is astonishing how little discussion and expertise there is in the IT field on the economics of information capital, and how little understanding there is of its cost dynamics. All the attention being paid to how to make the business case for IT investments and how to measure its business value has yielded few firm conclusions. IS specialists are poorly prepared for their business responsibility. Their education contains very little on the subject of the practical economics of IT development and operations. Not one of the leading textbooks on information systems contains even a single chapter on the topic. What discussion there is of economics looks mainly at the declining cost of hardware components, which only adds to senior managers' puzzlement and frustration—if the cost of hardware is declining 30 percent per year, why is the overall IT budget going up?

The literature extolling competitive advantages through IT almost never discusses economics. The "almost never" is based on an anal-

ysis of more than twenty books in the field. My own book on telecommunications, published in 1986 and updated in 1988, badly ne-
glects this topic. It, like the books of many of my colleagues, focused
on market opportunities, new products, and revenues, working
from the assumption that benefits would outweigh the costs of strategic initiatives. Because even firms that did gain a visible competitive edge through IT did not know their costs and measurable benefits, writers and teachers could not incorporate those aspects in their
storytelling.

The message to managers has been: "Don't worry about the costs.
They are worth the risk. We can't measure the benefits of strategic
innovations and infrastructures anyway, so don't get trapped into
naive cost justification."

Senior executives must and do worry about costs. This does not
mean that they do not acknowledge the value of a competitive advantage, but, after all, they could get an immediate competitive
edge just by cutting prices by 95 percent. Senior executives worry
that IT advantages come at too high a risk, too high a price, and
with hidden costs. The history of data processing in the 1960s and
1970s, when systems development cost overruns were the norm
rather than the exception, is cause for reasonable concern. So, too,
are the overblown promises of the office or factory of the future and
the cashless and paperless society.

The IS profession and the researchers, educators, vendors, and
consultants who work closely with IS managers need to move the
subject of the economics of information capital to the top of their
agendas. The credibility of Information Services units in the 1990s
relies on their doing so. IS leaders in the 1980s were those who
brought a business focus to their firm's IT planning. The IS leaders
of the 1990s will be those who add economic focus to business
relevance.

For much of the 1980s top managers in effect gave IS a blank
check. Awareness of the new link between IT and competitive advantage led management to release the brakes on IT investment.
Competitive necessity drove bankers to sink money into ATMs,
airline executives to invest in CRS, manufacturing management to
buy into CAD, CAM, and CIM, and distribution executives to invest
in customer-supplier links. No one cost justifies fire extinguishers
when the office is on fire.

But the blank check has been withdrawn. IT budget growth is
being slowed and services are being outsourced. There are both
constructive and destructive reasons for this. It is easy to point to

the destructive ones and for IS managers, consultants, and writers to continue to use them as excuses.

Many commentators have cited short-term thinking and obsession with quarterly earnings as major weaknesses of American business. IT, meant for the long term, is, like R&D, a reduction of today's profits to create a basis for tomorrow's. A second reason for short-term handling of IT is buyouts. For managers who trade companies, IT is mainly proven cost today and unproven benefits tomorrow. When the priority is to tidy up the balance sheet, there is almost no incentive to commit capital to IT. Third is cost justification. Managers who lack comfort with and good intuitions about IT will naturally retreat to their areas of strength, typically, short-term budgets and cost allocations. This is like cost-justifying R&D, management education, and pollution control. Fourth, firms' accounting systems are generally poorly positioned to track IT lifecycle costs—relating operations and maintenance to development, for example. Naive charge-out policies distort decisions about funding infrastructures. Finally, career interests and politics provide few incentives for managers to invest in expensive and complex IT infrastructures in organizations that do not view IT as a key business resource at the policy level.

What if all these problems disappeared? Would the issues and uncertainties of IS costs, business justification, and business value disappear? Surely not. Senior executives have many constructive concerns about the economics of IT. Business value *is* unproven. IS managers and planners generally do *not* have a firm grasp on long-term costs. It is ironic that we can predict fairly reliably the price five years hence of a unit of computing power or a fiber optic transmission link, but we cannot tell firms what they will be spending on IT overall or how to get the best value from that spending.

Aligning the Business/Technology Dialogue

The starting point for aligning the business/technology dialogue here is a floor-to-ceiling analysis of the cost side of IT. Building the IT Asset Balance Sheet described earlier is an immensely valuable first step because it starkly and directly repositions the discussion of the economics of IT away from the standard questions of "Why does IT cost so much?" and "Where can we cut costs?" to "How do we as managers take responsibility for this capital asset?"

The IT Asset Balance Sheet can help top management recognize that IT warrants management processes for an asset, not an expense. Software development and data, because they are not legally recognized as capital assets for accounting purposes, do not appear on the official balance sheet. When they turn out to account for a substantive fraction of a firm's real asset base, responsible senior executives see at once that IT demands a high-level corporate capital planning perspective and that strategies for funding development and building infrastructures cannot be treated as a case-by-case, bottom-up aggregation of applications.

Building the Asset Balance Sheet also helps senior executives to see that they have a responsibility to match level of authority with the business and economic importance, not of IT, but of IT capital. It helps them to see that, at the very least, the top management team must monitor the capital implications of IT and accord it as much time and attention in key planning meetings as any comparable asset base. Finally, it helps senior managers to see that they need to rethink the role of corporate Information Services management and business-unit IS management, particularly concerning responsibility for managing benefits and costs.

The IT Asset Balance Sheet can also alert senior executives to the difference between funding long-term platform infrastructures and individual applications. It raises a key issue that already dominates the IS management field: the balance between central coordination of platform-related areas of IT, especially the communications infrastructure, and decentralized exploitation of IT. The unsolved and often unsolvable problem of justifying infrastructures and identifying their business payoff falls to top-level corporate policy and planning. The vital need to ensure that IT developments respond to demands of competitive necessity calls for embedding responsibility for all aspects of IT in the business unit. The challenge is to see that platform and application needs move together rather than in conflict.

Finally, managers who study the IT Asset Balance Sheet come to realize that what is important is understanding cost dynamics, not costs. The capital perspective, particularly the capital impact of individual software development, alerts managers to just how much of the budget cannot in practice be cut and how much of the costs are hidden. The 1:4 relationship between development expenditure and capital commitment, the 50 percent of hardware and software capital costs that lie outside the corporate IS base, and the growing organizational support and education costs implied in individual expenditures are just a few examples.

Turning the integration monologue and the economics monologue into dialogue will not generate immediate and easy answers. There are no simple formulas for measuring the business value of IT, for determining cost and pricing strategies, defining and implementing the IT platform, or creating an integration path out of multivendor chaos. But just initiating the dialogue is a counter to the momentum of the historical IT planning process, which has emphasized applications over integration and budgets and expenditures over capital asset planning.

The Needed Dialogues

Dialogue is needed most right at the top of the firm. It is no exaggeration to say that nothing will contribute more to a firm's ability to take charge of change related to or fueled by IT than to have the firm's business and IS leaders make the issues of economics and integration a mutual priority. Doing so would serve to align the thinking of the two parties who determine the very basics of business choices and technology consequences and technology choices and business consequences.

When the economics and platform dialogue is well under way, other senior management dialogues just as essential for fueling progress in business design through information technology can begin. These dialogues are between the CXOs: the chief information officer, chief financial officer, chief human resource officer, and so forth across all business functions.

Each element of business design can be expected to pose difficulties. Organizational issues of human capital redeployment, for example, are likely to strain many firms' resources and skills. Many companies are poorly positioned to adapt quickly to the new global playing field, let alone mesh international business and IT planning and implementation. Many have only just begun to examine the implications of telecommunications for rethinking aspects of organizational design such as fields/headquarters linkages, team technologies, and location-independent operations.

However complex the individual issues, it is their sum that amounts to the tactics of using IT as a business resource. The need is to ensure continuing alignment among the six elements of business design and to end the compartmentalization of thinking and discussion among those responsible for the firm's business health and those whose formal charge is technical planning, implementation, and operations.

The Information Technology—Human Resource Link

One key area for seeking alignment through new dialogues is in the links between IT and organizational and human resource issues. For many companies, investing time and attention here may provide the largest single improvement in what may be termed *implementation productivity:* the degree to which the development, installation, use, and impact of any new system is leveraged, rather than blocked, by how well "people" issues are anticipated and handled.

Information Services units recognize that dramatically improving software productivity is one of their major responsibilities for the 1990s. Computer-aided software engineering (CASE) tools and object-oriented programming (OOPS) are seen as key building blocks for new disciplines in planning, analysis, design, programming, testing, maintenance, and communication with colleagues and clients.

It is too early to tell what impacts CASE and OOPS will have on lifecycle productivity. They represent as major a cultural change for IS professionals as office technology did for secretaries and managers. But, as so often with IT, for every claim of success and benefit there is a disaster or disappointment story.

CASE addresses the aspects of software productivity that are under the control of IS and contributes to those that are under the control of or at least influenced by human resource and business-unit managers, for example, real communication between designer and client, meaningful involvement and collaboration, and mutual understanding across the development cycle. Across the implementation cycle—for example, how change is introduced, how the dilemmas of work, jobs, and careers are managed, and how the business prepares, supports, educates, and rewards people—CASE can do little to improve effectiveness and productivity.

It is not the software but the human side of the implementation cycle that will block progress in seeing that delivered systems are used effectively. This was not the case in the 1970s, when taming the technology was the main challenge and just delivering systems the goal. As more and more IT-based business and organizational shifts are based on off-the-shelf software, and as CASE and related tools improve the speed, quality, and cost of development, the human side of IT will become ever more critical to success. This commonsense observation is highlighted by virtually every experienced practitioner and commentator. IS managers' commitment to communication, involvement, and service and to building new business skills is sincere.

But sincerity is no substitute for technique, and good intentions are no substitute for professional skills and organizational resources. Firms are now restructuring their entire organizations and processes, with and without IT. When IT is being used to position the firm for organizational advantage by redefining work at the same time that people are being "excessed," "outplaced," "delay-ered," and the like, the people side of IT will determine the pace, perception, and outcome of change. Taking charge of change here rests on human resource professionals being tightly linked to the entire IT planning process and developing a new depth of understanding of business design through information technology.

The quality of the relationship between the head of human resources and the head of IS can be among the most powerful factors in determining how easily and smoothly IT meshes into the organization. Quite often, the impact of this relationship is not apparent because it mainly removes likely disruptions and disharmonies. Nothing much happens visibly. Implementation remains difficult and challenging. People still moan about the costs, time, jargon, and delays always associated with IT.

What is happening invisibly, though, is that IS is buffered from its own likely blunders. The IS profession is not renowned for its insights on human relations. It is also not experienced in areas of education, job design, group processes, and introduction of change, all of which are now recognized components of its responsibilities. Ongoing dialogue between senior HR and IS managers ensures that someone qualified and skilled is seeing that education, careers, reward systems, job design, and the like are factored into IT and that the reverse is also occurring. There is convincing evidence that attitudes toward and expectations of IT strongly influence the outcomes of new systems and technology. Only when the leadership of the HR function is linked to the IT planning process and the thinking behind it can the expectations and attitudes be anticipated and addressed.

For the IS executive, the dialogue adds an ally and a professional. For the HR manager, it provides early alerts about where skills, jobs, recruitment, and organizational design can be expected to shift, either incrementally or dramatically, in coming years. For both, it adds mutual education.

Mutual education for HR and IS executives is not easy to achieve in other ways or through other means. Their respective staffs are seldom familiar with the organizational implications of IT. These executives need one another's experience and expertise for shaping active rather than reactive plans for taking charge of change and helping people across the organization feel comfortable with it. In organizations that lack this alliance, training departments that assume responsibility for IS education efforts are likely to sponsor programs that, because they are not tied into concerns about change or related to what is happening in the IS unit, are viewed as ineffective and irrelevant by attendees. The programs are detached from the context of implementation. Similarly, without this alliance, IS executives often have no strong adviser to help them reposition their own organization. They know that they need new organizational forms and that they must distribute some of their staff to the business units, to create new planning and support groups, to develop a service capability and, often, to totally restructure the central IS unit.

The Information Services function in most organizations is in a period of radical cultural change and associated discomfort. Its own HR needs are in many ways more urgent and stressful than its clients'. For all these reasons the HR/IS dialogue is essential.

A Missing Link—The CFO/IS Dialogue

The importance of the organizational and human side of IT is recognized, if not adequately addressed, in most firms. This is not the case in two other critical areas of IT planning and impact: accounting and finance. Here, history blocks progress. IT is a complex economic good and a major financial asset. It has been handled since its inception as a simple overhead item. H. Thomas Johnson and Robert Kaplan's influential book about managerial accounting in U.S. business, *Relevance Lost*, shows how accounting practices have significantly distorted planning and decision making.

The equivalent title for a similar book on information technology would be *Relevance Never There*. The typical accounting system for IT does not track lifecycle costs. It does not even indicate how much is being spent. Items are scattered across divisional and corporate budgets, with telecommunications, personal computing, and other expenditures appearing in a variety of categories. The system rarely provides a reliable indicator of the software asset base, or of relationships between development expenditures and maintenance and operations ratios. Both charge-out policies and methods for capital investment justification reflect the 1960s heritage of overhead allocation.

Information technology is capital, not overhead. It involves complex risks and trade-offs, and each of its many elements has its own cost dynamics. The interdependencies among them, the differences between IT infrastructures and applications, the problems and opportunities of integration path planning, the implications of particular choices of standards—all these "technical" issues are highly interlinked with financial issues. When a firm's financial accounting systems and traditions artificially and arbitrarily constrain economic options or misrepresent economic realities, as they so often do, appropriate decisions in terms of technology, business, and financial logic are penalized, and much less sensible ones prevail.

In the 1980s, business managers awoke to the new competitive relevance of IT. In the 1990s, managers in finance and accounting will wake up, not to the costs of IT, of which they are well aware, but to the need to go below the surface of line items and old accounting and planning procedures to find the costs and benefits. Too many persist in seeing their role as a guardian of orthodoxy, particularly with respect to charge-out. These managers are unwilling to adapt existing procedures to reflect the new nature of IT, its management, and uses.

Perhaps this resistance to change is a function of concern that IT might slip the leash of accounting and financial procedures, perhaps a basic unwillingness to come to grips with a difficult, jargon-loaded topic and deal with difficult, jargon-prone people who always seem to want more money and all too often have not delivered on their earlier promises. The role of financial managers is to ensure that the firm's money is responsibly invested and managed. If they continue to operate from a base of ignorance about IT cost dynamics and from a base of accounting systems and financial procedures built in the days when the firm had a single corporate computer center and a single systems development group, they are not meeting that responsibility. The same holds for IS managers.

The Information Executive: A Relationship, Not a Job

The role of the chief information officer, or CIO, has been much discussed in the business and Information Services press during the past few years. The CIO is seen as a new sort of information czar, someone with a broad mission and a strong business focus who will ensure that the firm's IT strategies meet its competitive needs. *BusinessWeek,* which in 1990 reported that CIOs were being fired at nearly twice the rate of 1988 and 50 percent more frequently than other senior executives, defined the term as now standing for "Career Is Over." This followed a period when CIOs were portrayed as the new heros. In January 1990, a few weeks before the *BusinessWeek* article appeared, a *Computerworld* article headlined "Million-Dollar Club Open to CIOs" observed that "senior IS executives, who were once considered little more than the mechanics of the information age, are now viewed as its architects."

There is no contradiction here between the CIO as both "endangered species" and "growing elite." This is the almost inevitable result of top management awareness that leads to delegation; firms upgrade the level of the job of spearheading IS—relabeling it Chief Information Officer—and expect the hero to perform, which *Computerworld* describes as sharing "the profit-making responsibilities traditionally shouldered by the chiefs of marketing, finance, and operations."

Expectations were high and unrealistic, particularly in the context *BusinessWeek* described:

> Facing heavy debt or lean times, many corporations are paring technology budgets. And CIOs, some of whom earn seven-figure incomes, are getting the ax. Others are hanging on, but with diminished power. The CIO title may sound impressive but it can be a hollow shell.

> More than just a technology whiz, the CIO was expected to think up strategically important information systems. . . . But despite this early rush to hire them, only a handful ever gained real power. The CIO is usually the odd man out among the executives. A half-dozen recent surveys say that less than 10% of CIOs take part in strategic planning sessions, and even fewer report to the CEO or president. Without influence at the top, it's impossible to make a difference. . . . And without a secure power base, even a visionary CIO is likely to lose the power struggle with other executives.

> When new information systems don't produce instant results, management tends to "blame the CIO."

> Still, embattled CIOs get little sympathy from their colleagues at other companies. Frequently, they say, chief information officers have been guilty of empire building—measuring their influence by the number

of computers and networks they control, not by whether they have helped to improve the company's competitiveness.[1]

The CIO fad obscured the difference between the job title and the relationship with top management needed to make the job effective. Delegation turns it into a fancy staff position with a big salary. If top management sees no reason to include the CIO in key business planning meetings, looks for instant results but expects long-term competitive payoffs, and believes all the stories about competitive advantage coming from waving a magic wand or using a consulting firm's 2 by 2 framework, disappointment and dismissal are bound to follow. Equally dismal are the prospects for the CIO who delivers little in the way of effective management of information capital or so overhypes IT as a source of competitive advantage that there is no hope of matching promise and reality.

Making the role of CIO as meaningful as the top management positions it imitates—CEO, CFO, and COO—requires new contributions by these other top managers, as well as by the CIO. Senior managers must commit time and attention to the competitive, organizational, and economic context of IT and include the head of Information Services on the top management team. Incumbents who lack the requisite skills and attitudes should be replaced. The job title does not make the manager; given the right person, backed by suitable authority and capable of earning influence, the title hardly matters.

Top managers' contributions to the relationship must include a willingness to face up to issues of IT architectures, the IT platform, and inherent lead times. First-rate Information Services managers most frequently attribute their inability to deliver results that they know are practical and significant to other senior managers' lack of understanding in these areas.

This book uses the terms "executive," "manager," and "head of Information Services" to identify the manager responsible for IT rather than CIO precisely because the situation *BusinessWeek* described in 1990 was obvious in 1985; the chief information officer's role is not a job but a dialogue, not determined by an individual but by a dyad. The CIO cannot make the job effective without the CEO and vice versa. Aligning business and technology begins here, with the close relationship of the top management team and the head of the Information Services function.

"Information Services" signals that this is a role that supports and enables business-unit activities. Chief information "officer" suggests far more independence from the wider business—an information czar instead of someone whose activities are highly interdependent with those of the business function. It must be remembered that the head of Information Services is supervising a technical resource that can be an enabler of business invention and innovation, a support for operations, and at the same time both a major cost element in doing business and a potential major revenue and profit contributor through partnership with the business.

If IS executives are to work in real partnership with senior business executives in marketing, finance, human resources, manufacturing, operations, sales, and every other function, the parties on both sides must understand the issues and their area of contribution, and see action as their responsibility and joint planning and action as a necessity. HR *must*, for example, accept responsibility for redeploying human capital. So, too, must IS. Finance *must* take responsibility for managing information capital in the terms described earlier. So, too, must IS.

CEOs can enable such partnerships, and realignment of business and technology by accepting that it makes sense to have a CIO only if they accord IS the same attention and commitment as finance. If they decide that IS is not a priority, they should neither hire a CIO nor expect any competitive contribution from IT. They might just as well continue to treat IT as overhead, invest only for reasons of competitive necessity, and handle IT investment as an expense rather than an asset.

For their part, IS managers must remove their blinders and take on the economics of IT. They may complain, with reason, about the short-term thinking, narrow focus on annual costs, and overdelegation that mark too many senior managers' handling of IT, but IT is, ultimately, more an economic than a technical issue. However much it affords in the way of competitive opportunity or however much it is deemed a competitive necessity, IT demands more and more capital that must come from somewhere else.

The most effective catalyst for real progress in IT is the combination of a top management team that understands the issues related to the IT platform and the architecture needed to position it and its own responsibilities for the platform at the level of policy, and a head of Information Services who fully understands the economics of information capital and his or her responsibility for them at the level of planning and oversight.

Aligning Core Business Drivers and
IT Springboard Initiatives

How might new dialogue between business and IT leadership set priorities for investment? It should go without saying that scarce and expensive capital committed to information technology instead of to some other area of the business must be targeted at high-payoff initiatives. Since the lead times for major IT developments are closer to those associated with R&D than with product marketing and manufacturing, this means looking ahead rather than waiting and reacting. The costs of reacting are apparent in the all too frequent gaps between expectations and outcomes of IT investments, and in the frenzied rush to bring in systems integrators to fix problems that a platform-centered approach to IT planning would have largely avoided.

Every large firm's use of IT today covers an immense range of applications. This range will expand rather than contract over the next decade. We cannot predict the 1990s' equivalents of desktop publishing, fiber optics, laptop computers, or neural computing, but we can be sure that 2001 will be as different from 1991 as 1991 has been from 1981.

How in this context can firms avoid the scattershot piloting of every new technology and application trend and scattershot choice of business investments in today's technology and applications? They cannot afford not to spend heavily on IT for reasons of competitive necessity; they cannot afford to spend heavily on it because of both scarcity of investment capital and shortage of development staff. And misspending has two expensive costs: the opportunity cost of either working on the wrong application or choosing the wrong technology, and the very real cost of wasted capital expenditure.

IT successes are rarely strategic in the traditional sense of grand and innovative. More often, they are conceptually undramatic initiatives that relate to the core business drivers of the firm and its industry and to the Braudel rule of changing the limits of the possible in the structures of everyday life. Aligning business priorities and technology investment depends on narrowing attention to the "springboard initiatives" for IT that support, extend, or differentiate core business drivers.

Information technology and change are mutually associated in business today and for tomorrow. IT brings business and organizational change and is fueled and paced by technological change. Any discussion of business and organizational change almost invariably highlights IT as a major contributor to it.

Change is what you think it is. Taking charge of change and treating it as an ally and opportunity rather than as a threat largely depend on building a sense of comfort about IT and its implications and on building the dialogues needed to mesh business and technology in the areas of planning, investment, design, use, and impacts. That is what this book has been about.

When senior managers recognize, collectively, that IT is embedded in business and organizational processes, that they are no longer managing the technology, but rather the context of the technology, they can begin to design business through IT. In the end, it will be the dialogue among these managers, not the technology, that will take charge of change and use it to create a competitive edge.

Shaping the Future: From Dialogue to Action

Dialogue must be followed by action. Any manager knows that there is no more business as usual. The 1980s changed the rules of competition and of organization. The 1990s are likely to make the past decade seem like an oasis of calm. After being taken by surprise or reacting with smugness in many instances, American business has geared up to address the realities of change as the norm, globalization as the new playing field, and customer power and quality as the very basis of competition rather than an add-on.

In each of these areas, managers are rethinking old assumptions, questioning old traditions, and abandoning old processes. They are changing the language of business. Management lore of the 1970s drew heavily on such terms as "decision making," "forecasting," and "span of control." The words that drive the thinking of today's leaders are different. Because they are so frequently used, it is easy to overlook the degree to which they shift the entire focus of planning and organization. The new lore highlights "market-driven," "timeliness," "flexibility," "time-based competition," "total quality," "total customer satisfaction," and "business process redesign."

To make these words more than clichés, we need to embed the new thinking in the methodologies we use. Firms are carrying out floor-to-ceiling reassessments of their basic processes, and bringing into the discussion workers on the factory floor, salespeople, executives, and customers. The same approach is being used in organizational planning. Companies are using teams to evolve the team-based organization. They are re-examining the linkages between business plans and information technology and vice versa. Each linkage demands new dialogues and new methods for structuring and sharing analysis in order to create new action.

Any methodology for analyzing business processes must begin and end with the customer, because it is customer power that has reshaped the terms of competition. Never before has the informed consumer and corporate customer had so many choices. Deregulation guarantees choices, at new prices. Globalization creates new sources of supply, at new levels of quality. Productivity and technology fuel overcapacity, which fuels more competitive products and aggressive pricing. Ease of access to services and information about them expands customer options.

Too many traditional approaches to business analysis, especially in the area of information technology planning, are poorly fitted to this new thinking. Many begin with the firm and work forward to the customer, instead of starting with the customer. They look at "quality" as a dimension of a product, not as an assessment made by a customer. Fernando Flores and his collaborators have developed a new approach. They point out that

> Customer satisfaction is not the same as quality. Take McDonald's, for example. Their products are not known for exceptional quality, and have not received rewards for either nutrition or cuisine. However, they manage customer satisfaction religiously, and have produced a vast enterprise based on producing customer satisfaction. Quality is an assessment that we have met some defined standard, while customer satisfaction is the judgment that a customer has about a product or a service. A customer can still be dissatisfied after taking delivery of a product that ordinarily gets high marks for quality.[2]

Many examples support these assertions. Airlines report that customer satisfaction *increases* when they make an error but handle it well, compared with not having made an error in the first place! In December 1989, Toyota recalled the 8,000 Lexus cars it had sold because of *two* customer complaints about small defects. The company arranged to pick up the cars, cleaned them thoroughly, repaired all minor flaws, provided a thorough tune-up, and delivered them back to the owners. Is the quality in the car itself or in the relationship between the customer and the provider? Is it defined by a measurement, such as defects per thousand, or by a customer's own perception of satisfaction? Did Toyota have a product quality "problem" or a customer satisfaction triumph?

If customer satisfaction is *the* issue for business design, we need to look at every element of business processes, work, and the IT systems that support or enable them in terms of customer satisfaction. None of the most widely applied methodologies for information technology planning meets this criterion. IT planning typically focuses on the firm's portfolio of applications and its operational needs. It is not unusual for a major firm's information systems plan or a consultant's report to omit a section that addresses the firms' customers, customer satisfaction, or customer feedback. When the business is rethinking its language, assumptions, and traditions, so must Information Services. Otherwise, dialogue will lead to insight but not to building useful and usable systems.

Flores and his collaborators, who include leading figures in a number of business and IT fields, have created a technology called Business Design Technology (BDT) that provides a powerful tool for making customer satisfaction an integral part of the planning and design specifications for business design through IT.[3] It addresses many of the issues raised in earlier chapters of this book, particularly the Braudel rule for assessing options for competitive positioning through IT, which focuses on changes in the limits of the possible in the structures of customers' everyday life. Business Design Technology also provides guidelines for action in the areas of organizational redesign, organizational simplicity, and redeploying human capital reviewed in Chapters 5 and 6.

The core principle of BDT is that the basic building block of business processes is a four-part "action work flow" that can be represented as the loop shown in Figure 8-1. The same four-phase work flow describes the business at all levels of operation. Complex processes can be broken down into subactivities, each comprising the action work flow loop. This means that the simple structure is valid for understanding and improving the fundamental interaction between the business and its customers and suppliers, as well as for enhancing, streamlining, or eliminating the simplest and smallest of the recurrent interactions among employees and managers at all levels of the organization that make up what we broadly call their "work."

Figure 8-2, panel (A), shows the four phases. Panel (B) diagrams a mortgage lending process. In the first phase of the action work

FIGURE 8-1 **Business Design Technology—Basic Action Work Flows**

A. BASIC ACTION WORK FLOW

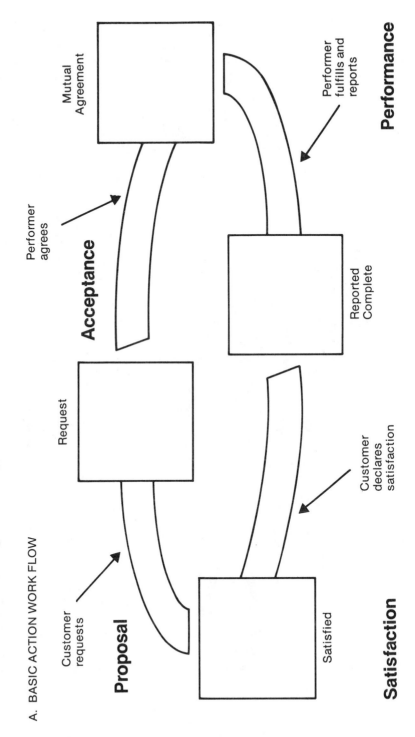

Proposal

Customer requests

Request

Acceptance

Performer agrees

Mutual Agreement

Performance

Performer fulfills and reports

Reported Complete

Satisfaction

Customer declares satisfaction

Satisfied

B. MORTGAGE LENDING MANAGEMENT STRUCTURE

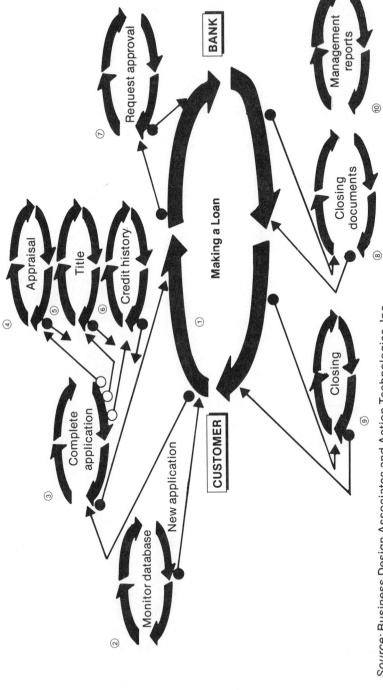

Source: Business Design Associates and Action Technologies, Inc. Copyright ©1990. Reprinted with permission.

flow, one party offers products or services to a potential customer (or the customer requests them). This proposal phase may include design, development, sales, and marketing activities, both before and during the sales situation. These activities may appear complex on the surface but in fact they all break down into the same construction of action work flows between "customers" and "performers" acting within the business itself.

In the second phase, acceptance, the "performer" in the work flow agrees to fulfill a specific condition in exchange for the "customer" meeting a corresponding condition. When the customer in the work flow is an external customer, the exchange is most usually a payment. This phase may include order processing and contract management, which in turn break down into a construction of action work flows.

The third phase is where performance takes place. Performance may include scheduling, production, assembly, and distribution. The final phase, satisfaction, closes the loop and involves the customer's acknowledgment or formal declaration of satisfaction. At its simplest, this is a "thank you" and payment. In more complex situations, it will include invoicing and payment processing, and sales representatives checking on delivery or following up sales and service with a client review.

All activities undertaken within the overall customer/performer loop obviously directly influence customer satisfaction, even if they are back-office or administrative work flows or work flows conducted between the business and its suppliers. An incorrect invoice for your "high-quality" toaster oven means a breakdown within the loop and leads to customer dissatisfaction. Every activity inside the customer/performer loop—and every information or processing system—can thus damage or enhance customer satisfaction.

As Figure 8-2 shows, complex business processes can be broken down into nested loops of the basic action work flow. The figure shows a business process design for mortgage lending. Many of the work flows encompass hand-offs from and to sales, production, and distribution staff. Breakdowns in coordination are common at such moments and thus affect customer satisfaction by creating delays, lost information, or misunderstandings. One of the distinctive strengths of the BDT approach from the perspective of business design through information technology is that it provides a base for targeting IT applications within and across the phases of the work flows. An automated teller machine is not carrying out a different process from that of a human teller. The customer request is,

"Please cash my check." It speeds up the entire customer satisfaction loop, reducing chances of errors and delays. The detailed subactivities look very different when handled by the ATM or a teller, but they follow exactly the same work flow.

ATMs and tellers each generate their own characteristic sources of customer satisfaction and dissatisfaction. For instance, the human teller can handle the agreement and performance phases of work flows more flexibly than most ATMs. Think of your own irritation when the ATM kicks out your card with the terse message, "Transaction not authorized. Contact your branch." Or, worse, keeps the card. The human teller can explain what happened: "I'm sorry, but there's a hold on your account. It was put on by the IRS. Would you like to talk with the branch manager?"

Future ATM designs based on assessing the customer satisfaction loop can be designed to provide new kinds of feedback and more sophisticated work flows than those in today's ATMs; for example, richer repertoires of messages, provision of customer service phone lines on a 24-hour basis, and new integrations of automated and human services. Just as action work flows begin and end with the customer and should be designed to close the loop with perfect customer satisfaction, IT systems features can and should be specifically targeted to contributing to that satisfaction and to warning when it is in danger or not present.

An additional value of BDT is that it provides the "primitives" (building blocks) for translating business process design specifications directly into software and information systems specifications. Complex activities such as mortgage lending or production scheduling can be decomposed into loops, phases, and subloops/subphases. There are clear linkages between BDT and CASE and object-oriented programming. BDT describes the work of both people and of machines in a general and comparable way. Here at last appears to be a base for translating business analyses *directly* to technical implementation.

Electronic data interchange, computer-integrated manufacturing, point of sale, image processing, telemarketing, on-line data bases, and every emerging high-payoff application of IT can be evaluated in relation to the phases of the basic action work flow. And, of course, Information Services organizations and professionals can stop talking about "users" and systems features and analyze their own services in terms of the loops that create customer satisfaction.

Fernando Flores has made major theoretical and practical contributions in many areas of management directly relevant to the topics of this book. His and his associates' development of the Business Design Technology is likely to become the most significant one. It fills in a missing element in the business design sequence that must move from business vision to policy to architecture to design to sustained action. Dialogue is the key to creating momentum for taking charge of change, but to turn dialogue into action we need simple ways of understanding business processes that are not simplistic. Above all, we need to make sure that IT planning, systems design, and implementation are carried out in a shared language and set of methods. Technical implementation is where business design through IT ends up. It begins with business vision, then moves to business policy. That provides the guidance for positioning the IT platform and defining the technical architecture. Vision to policy to architecture to design to implementation is the chain of dialogue and action.

I believe that what Flores and his colleagues have invented will be a foundation for the new discipline of the design of business and management that will emerge from the ongoing efforts of the best firms and their advisers to get ahead of the change curve. I think of this still-embryonic discipline as management chemistry, as compared with alchemy. Most of our management methods have evolved since the 1930s. Many are effective and built on experience, practice, and proven rules of thumb. They are alchemy, in the sense that they are not built on a fundamental set of elements and reliable principles. The medieval alchemists were not frauds or hucksters; they were skilled technicians, creating practical methods that worked within given bounds. I see in management thought and practice much effective alchemy. It works within given bounds.

Flores's work has been expanded into successful software products, papers, and courses. The core to his work is its focus on language: how we use language, and how we understand what we understand. This is ignored in virtually all of the discussions of communication, information, collaboration, and organizational structure. The brief notes I present here are sufficient only to hint at what I think will be a foundation of the design and management of businesses and of the information and communications systems that will support them.

I believe that we are positioning for as fundamental a shift in our understanding of management and design as marked the shift from alchemy to chemistry. I see in the best of the recent management literature not so much new answers as new questions, and a bold new willingness to rethink everything. We will still see fads, quasi-religious zealotry, simplistic panacea—alchemy—but the thinking of many of the writers that have been cited in this book promises a new discourse that will enrich our ability to create effective organizations that meet customers' wants and wishes across the globe and that enhance their members' well-being and growth.

Summary: The Effective Manager in the 1990s

I began this book by asking, "What will be the new skills required to be an effective manager in the 1990s?" The issue is management, not technology. Obviously, over the next decade we will see new and unanticipated business, technical, organizational, political, social, organizational and economic challenge, opportunity, stress, and uncertainty. Information technology will be a constant cause or correlate of much of these. There are no easy answers to the many questions they raise, and managers will continue to find IT a source of frustration and problems. They will also find it a source of excitement and solutions.

Today the IT field is at a pivot point. It comes out of a tradition of technocentered thinking, language, and methods and of poor mutual understanding between technical specialists and business managers. It is embedded in more and more areas of business operations but not yet embedded in the business management process. It has moved out of a long period when the technology was risky and too often did not work to a situation where we have more technology than we know how to use. It is shifting from a tradition of computing, where telecommunications was an add-on, to one of integrated technology platforms, where telecommunications provides the highway system into which computing applications fit.

Business managers are moving from a tradition where they could avoid, delegate, or ignore decisions about IT to one where they cannot create a marketing, product, international, organizational, or financial plan that does not involve such decisions.

At this pivot point, do *you* pull back, hold on, or swing forward?

Notes

1. Jeffrey Rothfeder and Lisa Driscoll, "CIO Is Starting to Stand for 'Career Is Over,' " *BusinessWeek*, February 26, 1990, pp. 78–80.

2. Quoted, with permission, from discussions with Fernando Flores and his collaborators. Their approach is described in a number of books, articles, and research notes. I have had the opportunity to review a number of unpublished papers prepared by the team for its own clients.

Recommended references from the team's published work are: T. Winograd and F. Flores, *Understanding Computers and Cognition* (Norwood, NJ: Ablex, 1986); R. Redenbaugh, "Beware the God of Quality," *Business Month* (June 1990); F. Flores and C. Bell, "A New Understanding of Managerial Work Improves Systems Design," *Computer Technology Review* (Fall 1984).

3. The methodology and technology are summarized in F. Flores and R. Dunham, "Business Design Technology" (Emeryville, CA: Business Design Associates, Inc.). In September 1990, Action Technology Inc. and IBM announced an agreement to explore jointly emerging technologies for use in managing business processes throughout their organizations. This is an early example of the fusing of methodologies for redesigning business processes and designing information and communications systems and software that can be expected to be major thrusts in business design through information technology.

A Selected Glossary
of IT Terms for Managers

Architecture

A central concept in the organizational evolution of IT, architecture is the technical blueprint for evolving a corporate resource that can be shared with many users and services. Because it defines how technical components will fit together, architecture involves a consideration of *standards*. The alternative to an architecture is to choose a technology base for each application based on criteria such as cost, efficiency, speed, and ease of use. This is a sensible strategy only if there is no expectation that applications will ever have to work together or share data or telecommunications resources, which is rarely the case today.

The technical details involved in defining an architecture are extraordinarily complex because of the huge variety of services and equipment it will be called upon to support. An effective architecture meets three often conflicting needs:

1) it provides as much vendor independence as is practical;
2) it rationalizes the multivendor, multitechnology chaos of incompatible elements that has evolved over the past decade; and
3) it provides a base for adopting the most widely accepted standards.

The firm that does not have an IT architecture does not have a real IT strategy. Left for another year, the problems of rationalizing a multivendor mess and moving toward integration, difficult enough today, will become even more difficult.

Automated Teller Machines (ATMs)

Taken for granted today as a basic access point for banking services, automated teller machines are representative of the competitive evolution of uses of IT in an industry. An ATM comprises three main elements: the workstation (the ATM itself); one or more remote computers that store customer records, carry out transactions, and control authorization and updating of accounts; and the telecommunications link between the ATM and the remote system. A wide range of technical and competitive options exists within and across these elements. A bank may build its own ATM network or share development and operational costs with a consortium of banks. It may design its systems to link into other services such as credit card providers' networks. The principal competitive issue that must be addressed in making these kinds of decisions is whether to lead or follow the leaders. The main technical issues relate to security, reliability, and efficiency. Customers perceive efficiency in terms of response time: how fast the ATM responds when a key is pressed. Adding a tenth of a second to each transaction can create growing delays in a network with 50,000 customers simultaneously trying to make withdrawals and deposits.

Backbone Network

The backbone network is the organization's central information highway system. It is analogous to the main airline routes, linking such cities as New York, Chicago, Atlanta, and Los Angeles. These cities are major hubs; smaller cities are linked to them. For example, traveling between two points in Virginia may require flying to Raleigh-Durham in North Carolina, rather than getting a direct flight.

In the same way, computer transactions and messages between such cities as Washington, DC, and Boston may be electronically hubbed through Chicago. Just as hubbing allows airlines to fly small aircraft into Atlanta, load passengers on Boeing 747s to Los Angeles, and then offload them onto more small aircraft, the backbone network provides the base for exploiting high-capacity telecommunications transmissions (747s) and advanced switching technology (airport security, air traffic control, and baggage handling facilities).

Bandwidth

Bandwidth is the carrying capacity of a telecommunications link. It determines the speed at which information can be transmitted, and consequently the practical range of applications the link can be used to provide. Bandwidth is usually expressed in terms of "bits per second," though technically it refers to the usable range of frequencies of the transmission signal—kilohertz, megahertz, and gigahertz—as for radio and television.

Microwave transmission rates are typically from 1.544 million bps to 45 million. Satellites use microwave.

An illustrative measure of bandwidth is the number of telephone calls that can be handled simultaneously by various telecommunications media.

"Twisted pair" telephone wire	4 to 240 phone calls
Coaxial cable	5,000
Satellite transponder	8,000 to 15,000
Optical fiber cable	15,000 to 20,000

Bar Code

The bar code is one of the most effective ways of capturing information about products. Portable scanning devices, laptop computers, and radio devices can quickly read bar codes and use the data as input to computer processing systems for, among other things, pricing at point of sale, updating inventory and delivery records, tracking physical movement of goods through the shipping cycle, and invoicing. This facilitates the development of just-in-time management alerting systems that provide up-to-date trend and exception analysis.

Bits and Bytes

Bits and bytes are nothing more than a coding system. Just as Morse code uses dots and dashes to represent letters and numbers, computers use zeros and ones. The entire computer and digital communications field rests on this "binary" 1-0 coding. Literally *any* information can be coded in this way, including numbers, characters, pictures, and telephone conversations. To the computer, each of these items is the same. The magnificent photographs of Jupiter and Saturn and their moons that were sent to earth by *Voyager* were digital. Compact disk music is digital.

Many aspects of computer and telecommunications capacity are measured in bits or bytes. A byte is simply a block of 8 bits that defines a single character, number, or other unit of information. Allowing for extra bits needed to indicate end of line, upper case, and so on, a reliable rule of thumb is that 10 bits approximate one printed character.

Approximate sizes in bits of messages and transactions are:

A credit authorization request	1,000 bits
An electronic mail message (1 page)	5,000
Digital voice (1 second)	56,000
Digital facsimile (1 page)	100,000
Full motion video (1 second, TV quality)	10,000,000

See also *Digital*.

Bridges, Routers, and Gateways

Bridges, routers, and gateways are progressively more complex devices used to link otherwise incompatible telecommunications networks. To a great extent, their use reflects lack of an architecture, permitting the proliferation of local area networks and personal computer facilities on a case-by-case basis with little or no regard to the almost inevitable eventual need to link them for business reasons.

Computer-Aided Software Design (CASE)

Computer-aided software design is a set of workstation-based software tools that let systems developers apply their own technology to their trade. Typical CASE software includes data dictionaries that store and validate definitions and define their use and calculation, diagnostic tools that check for inconsistencies and redundancies, and proven systems designs that can be reused in other applications and that allow report formats and screen displays to be designed at the workstation and diagrammatic representations of designs to be created quickly and kept up to date. Some CASE tools embody a particular development style or methodology based on "structured" methods and graphical representations of information and procedures.

Over time, CASE is likely to have the same impact on systems development as office technology has had on office work—and is likely to involve the same degree of organizational change and strain.

Compact Disk-Read Only Memory (CD-ROM)

Compact disk-read only memory is a laser technology, optical storage medium. A CD-ROM of the same size used in a record player can store more than 100,000 pages. Information can be retrieved in seconds via a standard personal computer.

Compatibility

Compatibility means fitting together. A piece of software and hardware are incompatible if they cannot be used together. Two operating systems are incompatible if software used on one cannot be used on the other, and so on through all the many elements of IT resources: computers, communications facilities, printers, software, data bases, and cables.

Incompatibility has been the norm in the IT field. Two types of standards are helping reduce it:

1) De facto standards created by the marketplace for users of IT
2) Stable standards implemented in real products by a plurality of key suppliers

It will be decades before incompatibility ceases to be a massive practical problem in using IT effectively. The primary goal of a firm's architecture is to move toward the integration that is the opposite of incompatibility and thus evolve a business platform for the delivery of IT services, sharing of information and resources, and the integration of business functions.

Connectivity

Connectivity is analogous to direct dial capabilities in international telephone communication. Telephone companies provide the connections, but unless the connected parties share a common language there can be no communication. Connectivity relates to the notion of reach provided and not the degree of range. See Chapter 7.

Cooperative Processing

Cooperative processing is closely allied with and extends the idea of distributed systems.

It is the notion of applications working together cooperatively, without respect to location or system. For example, a workstation might initiate a transaction and the software, recognizing that it needs information stored elsewhere, would retrieve it from the appropriate remote computer. Cooperative processing, because it permits a "Lego block" approach to building systems and products, is changing many aspects of systems and business product development.

Data Base Management Systems (DBMSs)

Data base management systems are information libraries. Individual applications use a master data dictionary to request the data they need. With a DBMS, data elements are entered and changes are made in one place, and audit, security, format definition, error-checking, and so forth are provided.

Attempts to create central, universal, single corporate data bases are giving way in many firms to efforts to evolve sets of cross-linked key data bases, clean up inaccurate data, generate consistent definitions, install new and reliable procedures, and extend relevant technology across more and more business functions.

Such "distributed relational data base management systems" make information a real and reusable asset. The technology for this is, however, inefficient and immature as yet.

Data Communications

Data communications was added to telephone systems designed to handle voice communications. Until the mid-1970s, a rough rule of thumb was that voice constituted 85 percent of a company's communications traffic and data 15 percent. Today, the proportions are being reversed, and digital transmission techniques better suited to data traffic are supplanting the analog techniques developed for voice.

Decision Support Systems (DSSs)

Decision support systems are information systems and analytical models designed specifically to help managers and professionals make more effective decisions. DSSs are not an independent technology or application, but tools that exploit any available and accessible technology to this end, often using personal computer software to access information in data bases throughout the organization.

Digital

The entire IT industry is built on an astonishingly simple base: the ability to represent any information as a combination of zeros and ones ("bits"), like an augmented Morse code of dots and dashes. The ones and zeros can be generated electrically as on-off signals.

The word "digital" refers to this 1-0/on-off/yes-no/true-false mode of representing, processing, and communicating information. It is contrasted to "analog." The old telephone system transmitted sounds by converting the sound wave to an analog(ous) electrical signal. Today's radio and television transmissions are similarly analog.

Digital communications offers massive advantages of cost, speed, accuracy, and reliability, which are essential for high-speed data communications. Voice communications benefits from digital transmission, too; it is just another form of information to be coded in bit form (like CD music).

For a few years, analog telephone systems and digital data communications will coexist, but more and more phone voice traffic will be shifted to digital communications. Just about all the world's major telephone systems will be fully digital by 1995.

Distributed Systems

Distributed systems link larger "host" computers with smaller, decentralized local computers. This contrasts with the pre-PC (or rather pre-intelligent workstation) era in which all processing was handled by large, centrally located computers, called mainframes, that interacted with computer terminals having no processing capability except through the connection to the mainframe.

Electronic Data Interchange (EDI)

Electronic data interchange is the electronic transfer of business information from one independent computer application to another, using agreed-upon standards for terms and formats of documents. EDI eliminates intermediate steps in transactions that formerly relied on paper transactions and information. Examples are placing purchase orders, sending invoices, and documenting freight shipments. The economic and organizational benefits that many firms report from using EDI, together with a trend among leading companies to choose suppliers on the basis of EDI capability, make this one of the major emerging competitive uses of IT for the 1990s. The most complex aspects of EDI are organizational rather than technical. EDI relies on rethinking business processes, not speeding up old bureaucratic procedures and paper chains. The two key standards for EDI are X.12 and EDIFACT.

Electronic Mail

Electronic mail directs messages to people, unlike telephone and facsimile, which send them to places. The principles of electronic mail are simple. The electronic mail service maintains a directory of subscribers and stores messages to these subscribers on disk. When a subscriber logs into the electronic mail facility—whether from home, office, or hotel—the mail system delivers all messages stored for that subscriber. Incompatibility among electronic mail services is being reduced by recent implementations of the X.400 standard. The explosive growth of portable fax is an indicator of the relative failure of electronic mail in terms of ease of use and compatibility.

Encryption

Encryption is the scrambling of information sent over telecommunications networks for security purposes.

Executive Information Systems (EIS)

Executive information systems are decision support systems targeted at business managers. Some present predefined and screened information, such as key market indicators or performance measures, but more and more pull data directly from firms' on-line business processing systems. Their purpose is to organize, analyze, and display information in a way that helps managers to get a clearer picture of key trends and events before it is too late to do anything about them. They represent a move away from management information systems based on historical accounting data toward management alerting systems built on up-to-date operations data.

Fiber Optics

A breakthrough in IT, fiber optics exploits laser technology to provide massive bandwidth; a single, almost invisible strand of optical fiber can carry thousands of telephone calls and transmit the equivalent of the entire holdings of the Library of Congress in a few seconds. Fiber speeds already exceed a *billion* bits a second.

Many telephone companies are currently replacing older copper cable with fiber. Compare with *Satellite*.

Image Technology

Image technology handles documents and pictures in much the same way that data bases handle purchasing records, ATM data, accounting information, and so forth. Images are captured in digital form by optical scanners, after which they can be stored, distributed, accessed, and processed much as any other data. Given Exxon's calculation that every document in its head office is copied 40 times and 15 of those copies are permanently stored in filing cabinets, and a Wang study that suggests that only about 5 percent of a company's information is stored in computers, the business and organizational opportunities of image technology are clearly immense.

Incompatibility (see Compatibility)

Integration

Integration is the process of making separate components of a technology base or business service work together and share resources. See also *Architecture*.

Interface

In the IT field, many aspects of standards relate to defining interfaces. A common interface enables dissimilar equipment and software to operate together. For example, given a precise statement of procedures, request formats, and electrical signals for transmitting data between a workstation and a local area network, hardware manufacturers, software developers, and makers of local area networks can design and implement their systems in many different ways, with different internal features. Provided all conform to the interface standard, they will be able to operate together. An effective architecture is built on standards that provide the maximum clarity about interfaces and the maximum freedom of choice about specific equipment and services.

Integrated Services Digital Network (ISDN)

Integrated Services Digital Network is the long-promised rebuilding of the world's telephone systems to support the combined transmission of all types of information—data, voice, image, and video—and the special facilities they can provide. ISDN is really a phone system for any type of information, delivered at speeds 10 to 30 times what is typical today. Additionally, ISDN allows one telephone line to handle two entirely different applications at the same time, managing a customer phone call as it manages the relevant transaction processing, for example. A 1970s' concept originating in Europe that is only now being implemented, ISDN is viewed by many experts and managers as obsolete and superfluous. Many of its design features, which were fairly advanced when defined, are already available in a variety of products, and the design has not kept pace with newer technologies and newer applications. Variations in the standards for implementing ISDN have not helped. It is unclear how effective ISDN will be in the United States. Businesses now have so many choices that they may not benefit from ISDN; consumers may, but not until the costs of monthly usage and the equipment that replaces the old telephone handset become attractive.

Local Area Networks (LANs)

Local area networks, which link computers and workstations within a single location, such as an office or building, evolved as a subfield of telecommunications. They help to satisfy the growing need for personal computers to send and receive messages, access data bases, share high-speed printers and high-capacity disk storage, and reach outside, via wide area networks (WANs), to other computers in widely dispersed locations. Early on, choices of local area networks were based primarily on departmental needs, price, and ease of installation and operation. Demands for business and technical integration have necessitated joint consideration of departmental and business needs. Incompatibilities between local area and backbone networks are a major problem in most firms today.

Mainframes

Computers have historically been classified in terms of their processing power, mainframes being the largest and minicomputers and microcomputers being progressively smaller. There is an additional category known as supercomputers, very-high-power computers with as yet limited relevance to business. Mainframes, long the workhorses of business computing, have traditionally been centrally located and connected to remote terminals and workstations via telecommunications. Although increasingly powerful personal computers and workstations can perform many functions more cheaply and efficiently than mainframes, large-scale corporate information management continues to demand large-scale information management technology—that is, mainframes. The term "host" is often used now instead of mainframe, together with the concept of the workstation being a "client" and a mainframe, mini, or micro being the "server" of data to the client.

Millions of Instructions per Second (MIPS)

MIPS is the IT equivalent of horsepower; it is a rough measure of the power of a computer. Like horsepower, it is an incomplete indicator that nevertheless helps to classify overall performance. With each new generation of computer boasting a higher MIPS rating, it has increasingly become a "gee whiz" type of term. MIPS are used mainly for comparing computers of a given type, such as large mainframes or standard personal computers. Cost per "mip" is a frequently applied measure of progress in the IT industry. In 1980, a mip cost around $1 million. In 1991, it costs less than $2,000.

Modem

A modem is a device that links a digital computer to an analog telephone system. Currently an essential element of personal computing, modems will disappear once the entire telecommunications infrastructure of the public telephone network is digital.

Network

A network is a telecommunications facility that links computers, personal computers, workstations, and other electronic devices. The concept of an IT platform presents two aspects of networks that are particularly relevant to business: *reach*, which defines the physical extent of the network, that is, which devices can connect to which; and *range*, which describes the extent to which the services and information accessible through the network can be usefully shared.

Network Management

Inasmuch as network failures are now business failures, network management is a strategic business issue. A complex electronic processing base may have many subscribers and a wide range of equipment and transmission facilities. Network management systems monitor these, providing diagnostics, alerts, and management reports, and taking action when any fail by rerouting traffic, for example. These systems, and the standards for them, are still embryonic.

Skills in network management and automated network systems are a vital investment, and the designs of the firms' technical architectures need to embed network management "end-to-end." Only a fraction of firms do this today.

On Line

On line means directly accessible from a computer or workstation. Off-line data, such as a magnetic tape or floppy disk, has to be loaded onto a tape drive or inserted into a personal computer before the program or service needing it can respond. On-line information is expensive and hard to manage but essential to customer service, just-in-time operations, and customer-to-supplier links. More and more of the costs of IT relate to managing on-line data than to "computing."

Open Systems

The notion of open, or vendor-independent, systems is one of the key ideas and battlegrounds of the late 1980s and continues to be an area of uncertainty and debate. A "proprietary" system may become effectively open if enough other vendors introduce products that are compatible with it. MS.DOS is an example. This operating system became the base for the personal computer industry, through "IBM compatibility."

Open systems will come about through a combination of users evolving de facto standards and the decisions of committees and vendors regarding which standards to follow. The OSI (Open Systems Interconnection) model is the international blueprint for open systems and has now been adopted by every major IT vendor as a substitute for a compatible complement to their proprietary systems. It will take at least another decade before OSI is fully implemented.

Operating Systems

An operating system is the software that manages the operation of the computer and determines which applications and software will operate on it. Operating systems have always been highly "proprietary," specific to particular vendors and even particular products sold by a vendor. They are the most complex piece of software that runs on a computer, far more complex than the business applications they supervise and manage.

Optical Fiber

See *Fiber Optics.*

Open Systems Interconnection (OSI)

OSI is the principal framework for implementing open systems. It is a blueprint that specifies not *how* systems should interconnect, only the interface for them to do so. Only parts of OSI are fully implemented as yet. It has provided a major impetus for progress toward reducing incompatibilities to the extent that IBM, regarded by many proponents of OSI as the black dragon of proprietariness, now offers several hundred products that comply with OSI.

There is no OSI tooth fairy who will magically provide comprehensive, instant, and proven open systems. OSI is to some extent a constantly moving target, as it needs to accommodate other standards and developments in local area networks, fiber optics, image processing, network management, and so forth.

Package

A software package is a set of programs that perform a specific function or set of functions in a generic way that meets the needs of a broad base of users. Packaged software varies widely in quality and requires as much planning as customized software development.

Personal Computer

A personal computer is designed for use by a single person. As PCs are becoming at once smaller and smaller and more and more powerful, the need to integrate them in networks and link them to larger computers and other networks of smaller computers is growing apace. Laptop and notebook computers, weighing 4 to 15 pounds, are the fastest-growing segment of the PC market. PCs are largely differentiated in terms of the operating systems they use, with MS.DOS, the system of the IBM-compatible machines, dominating the market. Apple's Macintosh systems are their main rival. IBM's OS/2 and Microsoft Windows 3.0 are the leading contenders for the corporate IT architecture.

Point of Sale (POS)

Point of sale, a simple concept with immense business implications, refers to the capture at the cash register (i.e., the "point of sale") of information that can subsequently be sent through the firm's network to update inventory systems. The transaction may also be linked to others at the same workstation. Point of sale is a selling opportunity and a basis for linking sales transactions to planning, ordering, pricing, inventory management, and so forth. It permits the evolution of real-time business operations. Electronic funds transfer at point of sale (EFTPOS) adds immediate payment to the POS transaction.

Private Network

A private network is a fixed-cost service with guaranteed levels of capacity and performance leased from a telecommunications provider, such as AT&T, MCI, or US Sprint in the United States, British Telecom or Mercury in Great Britain, and PTTs (Postes Télégraphique et Téléphonique) elsewhere. A company with the requisite skills can manage the traffic on a private network to maximize the rate of throughput at the fixed cost and can achieve greater security than is possible on a public network.

Virtual private networks, which offer cost advantages for large firms, ensure a contracted level of service without providing "dedicated" facilities. They offer new levels of flexibility, ease of expansion, and cost efficiency.

Proprietary

Vendor specific. See *Open Systems*.

Protocol

A protocol is a procedure for establishing a telecommunications link and exchanging information. Much of the incompatibility between telecommunications facilities is a function of their use of different protocols. Open standards, gateways, bridges, and protocol converters address this incompatibility in different ways and with different degrees of effectiveness.

Postes Télégraphique et Téléphonique (PTTs)

Postes Télégraphique et Téléphonique are quasi-government agencies responsible for a nation's telecommunications. Leading PTTs are British Telecom (an anomaly because it is part private company and part PTT), France Telecom (a firm monopolist), the Deutsche Bundespost (a recently reformed monopolist), and Japan's domestic NTT and KDD. The United States is the only country that does not have a PTT. Many PTTs are relaxing regulation in response to global trends in technology and business. They remain, however, a frequent blockage to large firms wishing to create transnational telecommunications services.

Public Data Networks

Public data networks are subscription-based, volume-sensitive telecommunications systems. Users pay as they go; the more traffic they send over the network, the more they pay. Contrast with *Private Network*.

Relational Data Base

A relational data base is organized so that elements stored in it can be cross-referenced. This deceptively simple concept is immensely complex technically and organizationally, involving entirely new types of software and heavy overhead in processing and operations.

Response Time

Response time is the time it takes an on-line service to respond to a user message or request. The quality of a system from a user perspective is strongly dependent on response time. Network managers and designers of on-line services set response time as a service measure, aiming at X seconds response time in Y percent of the cases, typically, 3–5 seconds in 95 percent of transactions. Heavy traffic on a network or a large number of transactions accessing the same software or data base can easily degrade response time.

Satellites

Communications satellites move information using very-high-frequency radio signals. Signals are beamed to a satellite's transponders, which amplify them and beam them back down to any receiver antenna in its "footprint," that is, the area of the earth covered by its broadcasting signal. Satellite transmission is not as cheap or fast as fiber optics but since it does not, except at each terminus, rely on cabling, it is ideal for countries such as India and Indonesia, where cabling would be prohibitively expensive because of the wide distances between key locations.

Smart Cards

A smart card looks like a credit card with an embedded computer chip. To date, smart cards have seen only limited application, but many commentators see a wide range of opportunity for devices that contain enough memory to store records and handle transactions without recourse to telecommunications links.

Systems Network Architecture (SNA)

Systems Network Architecture is IBM's main telecommunications architecture and the base for many *Fortune* 1000 companies' architectures. Historically, it has been seen as the proprietary alternative to open standards and OSI. Today SNA accommodates most elements of OSI, and other vendors, whether adopting OSI or their own proprietary architectures, accommodate SNA.

Standards

Standards are established in the interest of making it possible for products developed by different vendors, or multiple products developed by the same vendor, to function together. They have no legal weight; they are voluntary and unenforceable.

The American National Standards Institute (ANSI), which has been particularly influential in the standardization of programming languages, and its international equivalents, the International Standards Organization (ISO) and Consultative Committee for International Telephony and Telegraph (CCITT), have historically dominated the standards setting process. But as the pace of technological change outstrips the ability of such committee-based organizations to keep up, computer and telecommunications vendors and leading companies and industry groups are playing an increasing role both in defining standards and in influencing which will work in the marketplace.

Business has a strong interest in standards, yet is weakly represented in much of the standards-setting process. Managers in many leading U.S. companies are unfamiliar with both the standards-setting process and the industry user groups that cooperate with and try to influence them.

Systems Integration

Systems integration is the profession of making incompatible elements of a firm's IT base work together, developing plans for a higher overall level of integration to meet the firm's corporate business needs, and managing telecommunications, hardware, and software integration in developing new systems. One of the more recent additions to IT vocabulary, systems integration has been necessitated by the widespread technical incompatibilities in many firms that have become an operational nightmare and a business liability.

Telecommunications

Telecommunications refers to the transmission of information between remote locations.

Transmission

Transmission refers to the movement of information through a telecommunications network. It is concerned with establishing links, sending information, and ensuring that it arrives accurately and reliably.

UNIX

UNIX is an operating system that is highly "portable" across different operating environments and hardware and that supports "multitasking" (i.e., running a number of programs at the same time). Long a favorite of technical experts, UNIX is becoming popular in engineering and manufacturing circles. With IBM's introduction of a powerful set of UNIX workstations and leading UNIX providers' development of linkages with IBM's principal operating systems and telecommunications architectures, UNIX is becoming positioned to become a part of companies' architectures.

Value-Added Networks (VANs)

Value-added networks add something of value to the basic transmission provided by a telecommunications supplier, for example, electronic data interchange services, electronic mail, or information services.

VANs are more widely used in Europe than in the United States, largely because industries there have traditionally worked closely together to share resources, whereas large U.S. companies have tended to focus more on building "private" facilities. The growing importance of intra-industry and intercompany electronic transactions is making VANs an attractive option, especially in the international area.

Videotext

Videotext is a broad term for two-way interactive services between an information provider and standard personal computers or specially provided workstations. Intended to bring home shopping, home banking, newspapers, and masses of information into the home, videotext has met with little success, France Telecom's Minitel being the notable exception. The main problems with videotext have been cumbersome and slow communications and a lack of interest among consumers in the services it offers. The Prodigy service in the United States is the most recent effort to create a consumer market for videotext.

Videoconference

A videoconference is an electronic meeting. The main requirements for organizational videoconferencing capability are a sufficient number of rooms designed to send and receive video and a satellite "downlink" at each location. Today's cost of a room plus downlink, $50,000 to $250,000, is being reduced by improvements in earth receivers.

Very Small Aperture Terminals (VSATs)

VSAT technology, satellite telecommunications using very small earth stations that can be located and relocated quickly, is opening up many opportunities for linking locations that would be impossible if they required putting in cables, and is increasingly being adopted by large firms.

Wide Area Networks (WANs)

A wide area network links geographically separated locations. WANs involve transmission facilities far more complex than local area networks.

Workstation

Workstation is the most generally accepted term for the equivalent of a personal computer that is designed to operate fairly continuously in cooperation with other remote workstations, data resources, and transaction processing systems via telecommunications links.

X.25

The basic standard for international telecommunications, mainly because it is the one used by public data networks. It has particular technical characteristics that make it very suitable for some types of transmission and service, and much less suitable for others.

X.400

A badly needed recent standard, X.400 makes it practical for incompatible electronic mail systems to send and receive messages to and from each other. The standard covers all types of messaging systems, including e-mail, telex, and fax.

INDEX